+Q130 .W659 1991

Y0-CSM-293

Women in science

Q
130
W659
1991

DATE DUE

AUDREY COHEN COLLEGE LIBRARY
345 HUDSON STREET
NEW YORK, NY 10014

Women in Science

Edited by
Veronica Stolte-Heiskanen

For most of the twentieth century women have successfully embarked on scientific careers and made contributions to the advancement of knowledge in their chosen field. Although attitudes towards women pursuing scientific careers have obviously changed, gender equity in science is far from having been achieved. This volume is the result of a recent cross-national study, conducted by the European Co-ordination Centre for Research and Documentation in Social Sciences in co-operation with UNESCO's Division of Human Rights and Peace. It offers a survey of the situation of women in science in twelve European countries, with special emphasis on the obstacles and opportunities of access to positions of responsibility. Different regions of Europe are represented by studies from countries in all parts of the continent. A particular value of this volume lies in the fact that attention is paid also to women in smaller countries, about which our knowledge is very limited.

Veronica Stolte-Heiskanen is Professor of Sociology at the University of Tampere, Finland.

Women in Science

Token Women or Gender Equality?

edited by
Veronica Stolte-Heiskanen

and
Feride Acar
Nora Ananieva
Dorothea Gaudart
in collaboration with
Ruža Fürst-Dilić
for the
European Co-ordination Centre for Research
and Documentation in Social Sciences

BERG
Providence • Oxford

International Social Science Council
inco-operation with UNESCO

First published in 1991 by
Berg Publishers Limited
Editorial offices:
221 Waterman Street, Providence, RI 02906, U.S.A.
150 Cowley Road, Oxford OX4 1JJ, UK

© ISSC 1991, reprinted 1993

All rights reserved.
No part of this publication may be reproduced in any form or by any means without the written permission of the publishers

British Library Cataloguing in Publication Data
Women in science: token women or gender equality?
 1. Europe, Women scientists. Employment
 I. Stolte-Heiskanen, Veronica
 331.4815094
ISBN 0–85496–742–7

Library of Congress Cataloging in Publication Data
Women in science : token women or gender equality? / edited by
 Veronica Stolte-Heiskanen . . . [et al.] in collaboration with Ruža
 Fürst-Dilić for the European Coordination Centre for Research and
 Documentation in Social Sciences.
  p. cm.
 "International Social Science Council in co-operation with Unesco."
 Includes bibliographical references.
 ISBN 0–85496–742–7
 1. Women in science. I. Stolte-Heiskanen, Veronica, 1934–
II. Fürst-Dilić, Ruža. III. European Coordination Centre for Research and Documentation in Social Sciences.
Q130.W659 1991
305.43'5–dc20   90–28894
                                                              CIP

Printed in the United States by Integrated Book Technologies, Troy, NY.

Contents

List of Tables and Figures		vii
Preface		xi
Introduction *Veronica Stolte-Heiskanen*		1
1	The Emergence of Women into Research and Development in the Austrian Context *Dorothea Gaudart*	9
2	Handmaidens of the 'Knowledge Class'. Women in Science in Finland *Veronica Stolte-Heiskanen*	35
3	Women in Science Careers in the German Democratic Republic *Heidrun Radtke*	63
4	Double-Faced Marginalisation. Women in Science in Yugoslavia *Marina Blagojević*	75
5	Women and Science in Bulgaria: The Long Hurdle-Race *Nora Ananieva*	95
6	Soviet Women in Science *Vitalina Koval*	119
7	Women, Science and Politics in Greece: Three is a Crowd *Ann R. Cacoullos*	135
8	Women in Academic Science Careers in Turkey *Feride Acar*	147

Contents

9	Women at the Top in Science and Technology Fields. Profile of Women Academics at Dutch Universities *Esther K. Hicks*	173
10	Equal Opportunities for Women? Women in Science in Hungary *Agnes Haraszthy*	193
11	Stubbornness, Drudgery, Scientific Interests and Profound Commitment *Janni Nielsen and Bente Elkjaer*	199
12	Is to be an Engineer still a Masculine Career in Spain? Notes on an Ambiguous Change in University Technical Education *Maria Carme Alemany*	215
13	Recommendations *Dorothea Gaudart*	227
14	Select Bibliography: Women in Scientific and Technical Careers *Ruža Fürst-Dilić*	237
Notes on Contributors		249
Annexe A:	Other Specialists Who Participated in Various Stages of the Project	251
Index		252

List of Tables and Figures

Tables

1.1	Percentage of women by type and qualification of R & D personnel, by sector, 1985 (showing percentage change compared to 1981)	12
1.2	Percentage of women, by type and qualification of R & D personnel, by fields of science and technology, 1985, showing percentage change compared to 1981	15
1.3	Percentage of women scientists in universities and in other R & D-performing institutions, by fields of science and technology, 1985	17
1.4	Percentage of women in enterprise R & D by type of work and qualification, by branch of activity (ISIC), 1987 in full-time equivalent (FTE)	19
2.1	Women's participation in higher education, by field of studies, 1985/6	38
2.2	The share of women in research posts of the Academy of Finland	41
2.3	Percentage of women scientists at state research institutes	42
2.4	Percentage of women in positions of scientific power and prestige	48
2.5	Participation of women in the political and economic power structures of science and technology	52
3.1	The proportion of women in higher education according to fields of study, 1987	66
4.1	Women academics in institutions of higher education, 1986/7	83

List of Tables and Figures

4.2	The distribution of female university staff according to position, 1986/7	83
4.3	Women academics at institutions of higher education, by republics and provinces, 1986/7	84
4.4	Women academics (on faculties) according to field, 1986/7	85
4.5	Women in full-time jobs at scientific research organisations (and divisions), by republics and provinces, 1986	86
4.6	Women in full-time jobs at scientific research organisations (and divisions), according to field, 1986	87
6.1	The proportion of women among scientists (in thousands) in different positions, 1960–86	126
6.2	The structure of the scientific labour force in the USSR Academy of Sciences in 1988	127
7.1	Distribution of female students admitted in 1987/8 by field or department in three main universities	137
7.2	Science faculty by department or university school, rank and sex, 1985/6	141
8.1	Average time spent between two consecutive degrees/titles (years)	153
8.2	Distribution and share of women academics in different academic positions by fields, 1989	157
8.3	Distribution and share of women academics in different academic positions by type of university, 1989	161
8.4	Distribution and share of women academics in administrative assignments by type of university, 1989	162
8.5	Distribution and share of academics holding administrative appointment by marital status, 1989	167
9.1	Distribution, by field, of the sampled population and responses obtained	175
9.2	Total (full-time, tenured and untenured) professorial and lecturer positions at the three major technical universities in the Netherlands	180
10.1	The share of women graduates from higher educational institutions (at the beginning of the year)	194

10.2	Gender differences in scientific productivity among scientists in natural and technical sciences, 1980-5	196
12.1	Total number of students enrolled at the Higher Technical School for Engineers in Telecommunication, Polytechnic University of Cataluna	216
12.2	Students enrolled at University Schools of Technology in Spain	218
12.3	Total students enrolled in all years of study	218

Figures

2.1	Distribution of academic women by field and position in 1985	40
5.1	Distribution of women scholars by sciences and main subjects (comparative data in percentages)	101
5.2	Women academics in universities and higher institutes (comparative data in percentages 1978-87)	103
5.3	Women scholars by degree and academic position	104
5.4	Women doctoral students and women with higher degrees received from Bulgaria and abroad (percentages of total men and women)	105
5.5	Women doctoral students and women with higher degrees received from Bulgaria and abroad (by sciences, percentage of total men and women)	106
8.1	Proportion of women academics by field, 1989	148
8.2	Proportion of women with different academic status, 1989	149
8.3	Distribution of women into academic status categories by field, 1989	158
8.4	Proportion of women academics by type of university, 1989	159
8.5	Share of women in administrative positions by type of position and university, 1989	163
11.1	The most important qualities in a scientist	205
11.2	Factors influencing the choice of research theme	206
11.3	Attitudes towards women scientists'	

	qualifications (as compared to men)	209
11.4	Women's perceptions of male scientists	210
11.5	The influence of the structure and policies of academic institutions	211
12.1	Percentage of students having passed each year at the Higher Technical School for Engineers in Telecommunication, Polytechnic University of Cataluna	223
Appendix to Chapter 2	The academic teaching posts and research fellowships of the Academy of Finland (S.A.)	58

Preface

Women in Science: Token Women or Gender Equality? is the outcome of a cross-national research study project on 'Women's Participation in Positions of Responsibility in Careers of Sciences and Technology: Obstacles and Opportunities' conducted in 1988 and 1989 jointly by the International Social Science Council,[1] the European Co-ordination Centre for Research and Documentation in Social Sciences (Vienna Centre)[2] and UNESCO's Division of Human Rights and Peace[3] in co-operation with scholars, research institutes, academies of sciences and Unesco national commissions from East and West European countries.

Project activities began in 1988 and involved at different times researchers from a greater number of countries than are finally

---

1. The International Social Science Council was founded in October 1952, following a Resolution adopted at the VIth Unesco General Conference in 1951. the present ISSC is made up of fifteen international disciplinary organisations. Its major aim is to advance the social sciences and their applications to major contemporary problems by means of co-operation among social scientists and social sciences organisations at international and regional level.

2. The European Co-ordination Centre for Research and Documentation in Social Sciences (Vienna Centre) was created in 1963, following a resolution of the XIIth Unesco General Conference, and located in Vienna in accordance with an agreement between Unesco and the Austrian government. It is an autonomous body of the International Social Science Council and an international Non-Governmental Organisation (NGO). Its aims are twofold: to create and maintain co-operation between social scientists from East and West European countries and to stimulate cross-national studies in social sciences.

3. This project was launched pursuant to Unesco's Approved Programme and Budget 1988–1989, paragraph 13417 whereby the Division of Human Rights and Peace (Sector for Social and Human Sciences) was requested to undertake studies, in co-operation with research institutes and scientific NGOs, on obstacles to women's access to positions of responsibility in careers of sciences and technology. Some studies and reports produced by Unesco Sector for Natural Sciences and the Division of Statistics on Science and Technology likewise served as useful background information for the project.

Preface

represented in this volume. In some cases research was still in its initial phases. This volume brings together the work of social scientists in the research team from twelve countries: Austria, Bulgaria, Denmark, Finland, the German Democratic Republic (GDR),[4] Greece, Hungary, the Netherlands, Spain, Turkey, the USSR and Yugoslavia.

Three scientific meetings were organised during the course of the project to decide on its theoretical framework, to work out some common themes of investigation and subsequently to compare research findings. These meetings took place in Zagreb, Yugoslavia from 21 to 23 July 1988 in co-operation with the Association of Republican and Provincial Councils for Scientific Work of Yugoslavia; in Suzdal, USSR from 8 to 10 March 1989 in co-operation with the USSR Academy of Sciences; and in Lisbon, Portugal from 28 to 30 September 1989 in co-operation with the National Board for Scientific and Technological Research and the Portuguese National Commission for UNESCO. National Scientific Councils and Institutes contributed to the funding of research projects.

The present work constitutes an original and timely social science contribution to policy-oriented investigations concerning gender relations in the sciences. The studies help to shed light on the actual situation of women scientists and academics within their scientific workplaces and in their societies; what is the nature of the obstacles still impeding their career paths; why so many still tend to steer away from the 'harder sciences' (e.g. physics, engineering, mathematics); and to what extent they are able to play a role in defining the directions and content of research and experimental development (R & D) in science and technology. Women remain, on the whole, visibly a minority in prestigious academic and scientific bodies and have not yet gained an effective foothold in centres of decision for science and technology policies. However, many of the studies stress that women's creative potential and inputs can no longer be ignored in this fast-moving world where science and technology play a crucial role in determining the type and quality of life for present and future generations.

Taking into account the national specificities of R & D systems and underlying science policies on the one hand, and various

---

4. Research was conducted in the GDR before the re-unification of Germany.

Preface

levels of development of investigations of gender relations throughout European countries on the other, the individual studies vary in their levels of theoretical elaboration and the empirical research. Yet all have the special merit of placing the specific situations of women scientists within the wider context of R & D institutions and science and technology policies. So far social science research on gender relations and science has been uneven between countries in the European region. In several instances the country analyses are a pioneering first step in this area of enquiry.

Chapter thirteen, 'Recommendations', draws some practical conclusions based on research findings to which the attention of scientific establishments, governments and the UN system may be drawn.

The research papers included in this volume will be of interest and use to a wide range of audiences including the scientific community, policy-makers, the UN system and other international as well as regional bodies, non-governmental organisations and scientists themselves whether in the social, natural or physical sciences. We hope that they may serve as an inspiration to pursue a similar type of analysis in other countries and regions.

This publication is a result of collaborative efforts of the Vienna Centre, UNESCO and ISSC together with social scientists and research institutes from different European countries. The views expressed are, however, those of authors and therefore do not necessarily reflect those of UNESCO, ISSC or the Vienna Centre.

Introduction

Veronica Stolte-Heiskanen

About a century ago experts pronounced that in the course of evolution woman had been the 'loser in the intellectual race as regards acquisition, origination and judgement', and if 'an unfortunate female should happen to possess a lurking fondness for any special scientific pursuit she is careful to hide it as she would some deformity'.[1] Fortunately, the last hundred years have proved this claim, as well as many other scientific arguments of evolutionism, to be wrong. Everywhere women have successfully embarked on scientific careers and made contributions to the advancement of knowledge in their chosen field.

Although attitudes towards women pursuing scientific careers have obviously changed, gender equity in science is far from having been achieved. When it comes to positions of responsibility and power, gender inequalities are evident in practically all walks of professional life. However, the status of women in science is in particularly glaring contradiction with the scientific ethos that claims to follow the norm of universalism. Accordingly, science should be an endeavour wherein advancement and recognition depend entirely on individual merit. Yet, when it comes to women, practice still often falls far short of this ideal.

Science today has become essential to an ever-widening range of social, cultural, economic and technological activities, and it is an integral part of modern life for men and women. The hopes and expectations of economic and social benefits from science and technology are everywhere reflected in the growing efforts to increase the effectiveness, productivity and utilisation of

1. Kohlstedt, G.S., 'In from the periphery: American women in science, 1830–1880', *Signs*, vol. 4, no. 1, 1978, pp. 81–96.

scientific and technological research. In all European countries, national policies give high priority to the production of new knowledge and to the reproduction of qualified human capital, i.e. to research and training.

The decades following the Second World War witnessed the radical expansion and modernisation of the higher education and research systems in all European countries. Significant national differences notwithstanding, there has been a steady increase everywhere in the resources invested in scientific and technological activities, in the size of the qualified manpower potential, and science and technology policies have become an integral – sometimes even a strategic – part of overall national development policies of all countries. In view of the intensity of efforts on behalf of 'more and better science', optimal utilisation of the potential scientific and technological manpower should be one important goal. Science is one of the few remaining pursuits of mankind that depends very much on human labour: scientific knowledge is a product of the human mind – of both men and women. A pool of highly qualified women exists everywhere who could be encouraged to apply their knowledge and abilities to careers in science. Nevertheless, the prevailing spirit of rational planning and decision-making does not yet seem to extend to sufficiently active attempts to make use of the reserve army of qualified women. Empirical evidence presented in this volume as well as from other studies shows that women's opportunities to enter the institutional and organisational system of science and advance in it are considerably more limited than those of men.

Appropriate higher education is obviously a fundamental prerequisite for a scientific career. In this respect women have made dramatic progress over the last forty years. In numerous countries today as many women as men – often more – are enrolled at higher educational institutions and in some countries the proportion of women even among graduates exceeds that of men. Not only has higher education in general become accessible to women, but they are also increasingly entering the hitherto typically 'male fields' of the natural sciences and engineering and technology.[2] Yet, the radical progress in access to

---

2. For a recent review of women in engineering, see Michel, J., *Women in engineering education*. Paris: Unesco, 1988.

Introduction

higher education has not been matched by possibilities to pursue scientific careers on an equal footing with men.

The present volume surveys the situation of women in science in twelve European countries, with special emphasis on the obstacles and opportunities of access to positions of responsibility. Different regions of Europe are represented by studies from countries of eastern and western as well as northern and southern parts of the continent. Despite the variations in the cultural, political and socio-economic systems of the countries included there are also many similarities. In principle in all the countries women have equal rights on the labour market and in access to higher education. Practically everywhere female participation in the labour force has been steadily increasing over the last decades. Today women constitute from about one-third (Greece, the Netherlands, Spain, Turkey, Yugoslavia) to almost one-half (Bulgaria, Finland, German Democratic Republic (GDR), USSR) of the labour force. Also, the number of women enrolled at higher educational institutions has been continuously rising in all the countries. The rate of women's participation in higher education now ranges from about 40 to 45 per cent in Austria, Denmark and the Netherlands to 50 per cent or more in Bulgaria, Finland, GDR and the Soviet Union. While there are some differences in the relative distribution of women students according to fields of studies, on the whole a very similar pattern continues to persist everywhere. In general, the highest proportions of women students are in the humanities and the social and medical sciences, and the lowest in engineering and technology.

Women in all twelve countries are found in increasing numbers in professions requiring higher educational qualifications. Yet, everywhere the higher one goes up the ladder of the occupational status hierarchy, the fewer the women. The share of women in managerial and higher leadership positions has increased everywhere at a considerably lower rate than the growth of the number of potentially qualified women.

Science and technology are expected to play an important role in advancing the welfare of all the countries included in this volume. Scientific and technological activities are pursued with varying degrees of intensity, as is reflected in the share of R & D investments of the Gross National Product, which ranges from about 0.3 per cent to around 3 per cent. On the other hand, the

3

Introduction

growing importance accorded to science and technology is evidenced by the continuous increase in R & D investments in all of these countries. Similarly, with the exception of the Soviet Union, they are all relatively small countries on the fringes of contemporary big science centres of the world. Moreover, for linguistic reasons, scientists from Austria to Yugoslavia face similar problems of participation in the international scientific community.

As the selected bibliography at the end of this volume shows, during the last decades there has been a proliferation of research on the problems of science and gender. For a variety of well-known reasons most of the visible research results come from the mainstream of the world science centres. Consequently, relatively little is known by the international community about the situation of women in science in the smaller and/or linguistically isolated countries. The present collection of studies aims to fill this gap by making visible the world of women scientists in societies less familiar to the international public.

The scope of this volume ranges from comprehensive national overviews to specific cases of empirical research. The first four chapters present extensive analyses of the overall situation of women scientists in the respective countries, mainly based on secondary sources ranging from documentary materials and statistics to empirical studies on various aspects of women scientists' careers. Chapters five to ten, on the other hand, focus on women in academic environments within the broader societal context. In the case of the Eastern European countries this means women scientists employed by the Academies of Science, and analogously, in Western Europe, the professional staff of the universities. Chapter eleven presents the results of a survey carried out on a sample of academic women in the natural and technical sciences. Finally, chapter twelve is a comparative study of a sample of male and female engineering students at a Spanish university. Although the main concern of this volume is with professionally active women scientists, because of the universal scarcity of women in engineering and technology it was felt important to include also a study focusing on the training phase of careers in this field.

Variations in the scope of the chapters reflect differences in the degree of institutionalisation of research in this area in the respective countries. In Western Europe there is a relatively

Introduction

long-established tradition of research on women in science. Starting from the 1960s the Women's Movement – with predominantly professional membership – stimulated interest in the status and problems of professional women and their dual careers; the growing labour force participation of women led to the rise of interest in women by policy-oriented researchers; and critical re-evaluations of Freudian and Marxist theories led to formulations of theoretical approaches to the problems of women in a patriarchal society. These various trends also inspired a considerable body of empirical research on the problems of women with scientific careers.

In the Eastern European countries, on the other hand, in the 1950s and 1960s research interest in the 'question of women' was to a great extent guided by the then prevailing problems of the socio-demographic situation and the labour market. Consequently, most research focused on women's problems and advancement possibilities at work in general; on the relationship between work and family life and on the division of labour in the family and the attitudes of husbands and wives towards women's work outside the home. In the 1970s, attention began to turn to women in the professions (including to some extent scientific workers) and to social security provisions for working women. The relative lack of literature on women in science from Eastern Europe and the USSR included in this volume reflects the recent emergence of research interest in women's participation in positions of responsibility in science in these countries.[3]

Although no attempts were made to aim at systematic cross-national comparisons, certain patterns emerged out of the individual studies included here, which lead to some general conclusions concerning the obstacles to women's advancement in scientific careers. Many of the results found in previous studies of women scientists in other societies also find confirmation in each of the individual countries surveyed here.[4]

With depressing uniformity, irrespective of the country, type of research organisation, scientific discipline, or degree of

---

3. For regional surveys of trends in studies on women, see *Bibliographic guide to studies on the status of women*. London: Bowker/UNIPUB/Unesco, 1983.

4. For trends in research on women in science, see Stolte-Heiskanen, V. *Women's participation in positions of responsibility in careers of science and technology: obstacles and opportunities*. Department of Sociology and Social Psychology, University of Tampere, Working Papers, 26, 1988.

Introduction

'feminisation' of the field, the higher the status of the position held in the hierarchy, the fewer the women. Science represents no exception to this 'universal law' of women's fate.

At the beginning of this century the English psychologist James Swinburne proclaimed that 'the feminine mind is quite unscientific'. He did, however, consent that 'many women can do some sort of scientific work. They are more careful than men, and more accurate in taking readings. In this direction they make excellent assistants.'[5] The world of science apparently took Swinburne's advice to heart. With the interesting exception of Turkey, in none of the countries included in this volume is the share of women among professors higher than somewhat over ten per cent, and in many countries their proportion is even considerably less. On the other hand, women are increasingly making headway as 'assistants', occupying lower level positions in the academic hierarchy. However, on all levels the number of potentially qualified women substantially exceeds that of those actually engaged in scientific occupations. Moreover, in countries where some information is available, it seems that this hierarchical pattern is even more pronounced outside the academic sector, such as in mission-oriented research institutes and research units of the productive sector.

The importance of relevant role models for women's decisions to pursue scientific careers has been well documented in earlier studies.[6] The common pattern of few women professors or other senior scientists implies the lack of a needed critical mass to encourage and support women students on the one hand, and to speed up changes in the often noted gender culture of the scientific environment on the other. Women's lack of confidence coupled with exposure to the condescending attitudes of male colleagues is also a common finding in a number of studies reported here.

It also seems that, to a varying extent, a fundamental con-

---

5. Swinburne, J., 'Feminine mind worship', *Westminster Review*, no. 158, 1902, p. 189. Quoted by J. Sayers, 'Psychological sex differences', in *The Brighton Women and Science Group: Alice through the microscope. The power of science over women's lives*. London: Virago, 1980, pp. 42–61.

6. See e.g. Cole, J. and Zuckerman, H., 'Marriage and motherhood and research performance in science', *Scientific American*, no. 256, 1987, pp. 119–25; and Zuckerman, H., 'The role of the role model: the other side of the sociological coin', in O'Gorman, H. (ed)., *Surveying social life*. Middletown, Conn: Wesleyan University Press, 1988.

tradiction exists everywhere between the ideology and objective need for women's equal participation in science and the appropriate social conditions necessary for carrying out their professional roles on an equal basis. Thus, nowhere has the conflict between the biological clock, the domestic clock and the research system clock been satisfactorily resolved. Despite the existence of extensive social security support systems for women in many of the countries, the fact remains that the advancement structure of science has remained inflexible to the specific constraints faced by women, i.e. that the crucial period for ensuring a successful scientific career coincides with the childbearing and childrearing age.

While social and cultural factors can undoubtedly play an important role in contributing to women's career and advancement opportunities in science, not all the blame can be put on society or culture at large. All the studies also confirm the less frequently voiced fact that the scientific community as such is highly resistant to change.

Science is sometimes defined as a way of seeing things. Women still far too often constitute a 'blind spot' in the vision of the scientific community. The factors contributing to the invisibility of women in the scientific community are well documented in a number of studies from the present selection of countries. Variations in the position of women scientists in different scientific fields, in diverse types of research institutions, and between countries with varying degrees of intensity of overall R & D activities suggest that the more the sector represents an 'instrument of power', the less open its scientific community is to women. Resistance to the penetration of the power structures of science is reflected in the fact that everywhere women are notably absent from positions of higher levels of leadership and responsibility both within scientific institutions and in the intra- and extrascientific infrastructures concerned with science policy and administration.

Undoubtedly, women's position and advancement possibilities in science have improved everywhere over the last decades. There is also much room – especially at the top – for improvement in each country. Gender inequality in science is one instance of the denial of basic human rights. The overt or covert restriction of women's access to pursuing intellectual activities and professions of their choice – such as a scientific career – also

Introduction

implies unequal access to power. In view of the growing importance of science and technology in guiding societal planning and action in the contemporary world, sometimes characterised as the 'knowledge society', equal rights to scientific careers are of ever-increasing significance.

From a societal perspective, gender inequities in science represent an under-utilisation of a significant proportion of society's human capital. Since progress in the higher education of women has not been matched by possibilities to pursue scientific careers on an equal footing with their male colleagues, there is clearly an inefficient use of societal investments in highly trained professionals.

Finally, from the perspective of science, the unequal representation of women has implications for the production of scientific knowledge. The male dominance in science may not only have epistemological consequences, such as distortions of knowledge claims of science in general – as the feminist critiques of science suggest – but also results in substantive gaps in our knowledge of the world we inhabit. The invisibility of women in science in many disciplines leads to the absence of research interest and consequent lack of information on many problem areas of particular concern to women's lives. To what extent women can influence the direction of scientific activities greatly depends on their presence in positions of power and responsibility. Thus, the relative degree of women's access to positions of leadership and responsibility will have consequences not only for their status in science as such, but also for our future visions of society.

1

The Emergence of Women into Research and Development in the Austrian Context

Dorothea Gaudart

> *Equalizing of the gender-related imbalances in education is a tedious and long-term, however irreversible development process.*
> Hertha Firnberg, 1967

Introduction

The paper starts from the assumption that women, firstly, could only emerge into Research and Development (R & D) in the Austrian context when they had overcome the historically caused deficiencies in higher education, i.e. when they had increased substantially their access to, and graduation from educational institutions of the tertiary level. Secondly, research policy in general associated with high demands for labour by both the public and the private sectors affected women's participation in R & D. These trends are, however, also subject to national and international economic developments. Furthermore, it can be hypothesised that both the public and the scientific community underestimated the influences of international and national policies aiming to promote the status of women by increasing their participation in positions of responsibility in careers of sciences and technology, if they became aware of them at all. In this respect it is further assumed that the influence of international agenda setting on the one hand, and that of the women's movements and their engagement in women's studies on the other, will be decisive for accelerating future developments.

The Education System

Women gained access to higher education in Austria only 100 years ago. The fact that the production and dissemination of knowledge excluded women until then has determined in the following periods gender-specific disparities in participation in education, preferences for certain fields of studies, share of graduates and, consequently, in potential scientific and technical personnel. (Austrian Federal Ministry of Science and Research, 1981)

During the post-Second World War decades, one of the objectives of university reform was to improve the performance of Austrian science to catch up with international scientific developments. The participation of women in university and extramural research was not an explicit objective at that time. However, women took advantage of governmental policies to rectify inequalities in education. The expansion of higher secondary schools in the 1960s – with the objective of eliminating regional and social inequalities and under the slogan 'to each district its secondary school' – led to higher numbers of girls qualifying for university entrance in the 1970s. The proportion of women in first enrolments at tertiary level exceeded 50 per cent of all entrants in the mid-1980s.

This numerical gender balance in first enrolments in higher education did not, however, reflect the number of graduates. In 1970, around 25 per cent of all graduates were women. The figure rose to 39 per cent in 1984/5 and will reach 45 per cent around the year 2000. The percentage of female post-graduates is still lower (29 per cent).

In the Austrian resident population (aged fifteen years and more) the proportion of graduates from universities and equivalent institutions has steadily increased during the past thirty years. The percentage of male graduates increased from 3.1 in 1951 to 4.8 in 1981. The corresponding percentage for women increased from 0.5 in 1951 to 1.0 in 1971, and especially dramatically during the last decade to 2.3 in 1981. In the same year the proportion of women graduates in the 25–29 year old age group was 7.1 per cent, as compared to 6.6 per cent of that of men. (Austrian Central Statistical Office, 1986)

In conclusion, quantitative equality between female and male first enrolments in universities has been achieved today and a further increase of first entrants is expected. However, the graduation rate of women as compared to that of men will not yet be equal in the near future.

Employment Trends in Different Scientific Fields

Women took advantage of labour market policies aiming at higher levels of employment. Structural shifts in employment towards the tertiary economic sector contributed to an increase in women's proportion of the labour force, which included their involvement in R & D. Positive employment prospects for women in R & D personnel are therefore discernible when analysing gender breakdowns of the R & D survey data.[1]

Between 1981 and 1985 the proportion of women in the total personnel of R & D institutions rose from 34.5 to 34.9 per cent, however, clearly indicating a trend towards higher qualified staff. The proportion of women in the category of auxiliary personnel decreased, while it increased in the category of technicians and rose even more in the category of scientists and engineers.[2]

---

1. The structure of the Austrian scientific communities is periodically investigated according to the Unesco Statistical Questionnaire on Scientific Research and Experimental Development. The higher education and the general service sectors as well as the co-operative subsectors were investigated in 1969/70, 1975, 1981 and 1985 (gender-specifically since 1975) by the Austrian Central Statistical Office (R & D Division: Karl Messman, Elisabeth Scholtze). The business enterprise (productive/enterprise R & D) sector was surveyed in 1966, continuously in 1978, 1981, 1984 and 1987 (gender-specific since 1987) by the Austrian Federal Economic Chamber. A part of the R & D data published by the Central Statistical Office is included in the yearly Governmental Research Reports prepared for the Austrian National Assembly, however, without gender classification; latest edition: *Forschungsbericht 1989*, ed. Federal Ministry of Science and Research. Consequently, e.g. the *Reviews of National Science and Technology Policy, Austria* carried out recently by the Organisation for Economic Co-operation and Development (OECD) did not help to enrich the pool of available knowledge on gender-specific human resource management in R & D in Austria although this report notes the biggest increase being in the number of Austrian women students. *Reviews of National Science and Technology Policy, Austria*, OECD, Paris 1988, p. 121.

2. 'Technicians' refer to persons engaged in scientific and technical activities who have received vocational or technical training in any branch of knowledge or technology. 'Auxiliary personnel' refers to persons whose work is directly associated with the performance of S & T activities, i.e. clerical, secretarial and administrative personnel, skilled, semi-skilled and unskilled workers and all

Table 1.1: Percentage of women by type and qualification of R & D personnel, by sector, 1985 (showing percentage change compared to 1981)

Sectors	scientists/engineers			technicians			auxiliary personnel			total		
	Total (men and women)	%f	% change 81/85	Total (men and women)	%f	Total (men and women)	%f	Total (men and women)	%f	Total (men and women)	%f	% change 81/85
Higher Education Sector												
universities (excl. hospitals)	9 695	16.8	+9	2 516	69.8	3 147	65.4	15 358	35.4			
	7 510	15.9	+14	1 627	66.1	2 408	61.5	11 545	32.5	+3		
university hospitals	1 765	18.6	+2	718	86.6	642	78.3	3 125	46.5	−5		
art schools (university status)	129	22.5	−11	9	44.4	11	81.8	149	28.2	−5		
Academy of Sciences/Humanities	253	28.5	−2	135	36.3	65	86.2	453	39.1	−1		
higher technical schools, experimental stations	38	7.9	+75	27	22.2	21	38.1	86	19.8	+20		
Government Sector[a]	1 225	19.3	+12	833	32.2	2 683	43.3	4 741	35.2	+7		

Private Non-profit Sector[b]	582	27.1	+29	281	70.8	292	77.4	1 155	50.5	+5
Business Enterprise Sector (excl. enterprise R & D) co-operative	695	7.8	−13	654	21.9	1 488	34.3	2 837	25.0	−6
research institutes[c]	548	8.2	−12	553	22.1	1 311	34.6	2 412	25.7	−4
civil engineering and technical services	85	8.2	−14	47	36.2	65	50.8	197	28.9	+9
electricity-supply companies	62	3.2[d]	—[e]	54	7.4	112	22.3	228	13.6	−34
TOTAL	12 197	17.0	+10	4 284	55.2	7 610	52.0	24 091	34.9	+1

Source: Austrian Central Statistical Office, R & D Division.

a. Federal institutions; institutions of provinces, communities, chambers, social security authorities; museums; (without hospitals).
b. including institutes and research units of the Ludwig-Boltzmann Society.
c. including the Austrian Research Centre of Seibersdorf.
d. Percentage based on data from 1985; no gender breakdown was available in 1981.
e. Percentage cannot be calculated due to lack of necessary data for 1981.

The pattern of women's employment in R & D is similar to their general employment, i.e. higher numbers of women in the public sector (government sector in particular) than in the private business enterprise sector. The relatively highest proportion of women is economically active in the private non-profit sector.[3]

The low proportion of women among scientists in R & D-performing civil engineering and technical services (8.2 per cent) corresponds to the low number of female students. Consequently, the proportion of women scientists was below 10 per cent in the co-operative business enterprise sector (excluding the productive enterprise R & D in companies) and in the higher technical schools with experimental stations. However, in the latter the proportion of women increased from 4.5 per cent in 1981 to 7.9 per cent in 1985. In general, in the university and government sectors about one out of five scientists was a woman, while in the Austrian Academy of Sciences/Humanities and in the private non-profit sector, nearly one out of three scientists was a woman. During 1981 to 1985 the proportion of women scientists increased in the private non-profit sector and in the government sector, and slightly decreased in the Austrian Academy of Sciences/Humanities.

The distribution of R & D personnel by main fields of science and technology shows that the highest percentage increase of women in R & D institutions is in the social sciences and humanities. In each of these fields from the total R & D personnel two out of five are women; on the highest qualification level one out of five scientists is a woman.

There are more women scientists employed at universities than in the other R & D-performing institutions (in social sciences 274 compared to 98, in the humanities 410 compared to 232 and in the field of natural sciences 184 compared to 73).

Both universities and the other R & D units show increases of

---

other supporting personnel. Cf. *UNESCO Manual for Statistics on Scientific and Technological Activities*, ST-84/WS/12, Paris 1984.

3. The biggest institution in this private non-profit sector is the Ludwig Boltzmann Society – Association for the Promotion of Scientific Research, founded in 1960, as an 'umbrella' organisation for research institutes and centres, for the most part financed by the Federal Ministry of Science and Research. The Joanneum Research Society is a regional organisation of non-university research units in the province of Styria.

Table 1.2: Percentage of women, by type and qualification of R & D personnel, by fields of science and technology, 1985, showing percentage change compared to 1981

Fields of science and technology	scientists/engineers total (men and women)	%f	% change 81/85	technicians total (men and women)	%f	% change 81/85	auxiliary personnel total (men and women)	%f	% change 81/85	total (men and women)	%f	% change 81/85
A	2 540	10.1	−2	761	41.5	+14	1 040	53.1	−1	4 341	25.9	−3
B	2 080	4.4	+7	947	24.8	−3	2 036	34.9	−15	5 063	20.5	−1
C	2 897	20.8	+10	1 388	81.0	−3	1 198	74.0	−1	5 483	47.7	−1
D	640	17.2	+9	441	41.7	−2	1 245	49.5	−2	2 326	39.1	0
Subtotal	8 157	13.0	+6	3 537	52.6	0	5 519	50.1	−7	17 213	33.0	−3
E	1 756	21.2	+20	320	80.9	+10	497	83.1	+5	2 573	40.6	+8
F	2 284	28.1	+12	427	58.1	+4	1 594	48.9	+10	4 305	38.8	+11
Subtotal	4 040	25.1	+14	747	67.9	+6	2 091	57.1	+5	6 878	39.5	+10
Total	12 197	17.0	+10	4 284	55.2	+1	7 610	52.0	−4	24 091	34.9	+1

A = Natural sciences; B = Engineering and technology; C = Medical sciences; D = Agricultural sciences; E = Social sciences; F = Humanities.
Source: Austrian Central Statistics Office, R & D Division.

about 10 per cent for women scientists in medical sciences. In absolute figures nearly ten times more women scientists are economically active at universities (553 women out of 2,726, i.e. 20.3 per cent; in hospitals 328, representing 18.6 per cent) than in the other R & D-performing institutions. Between 1981 and 1985 there was also a 9.0 per cent increase in women scientists/ engineers in the agricultural sciences, again more at universities.

In line with the general employment pattern of women it is obvious that in the research sector also, many more women are economically active in R & D on the level of technicians and as auxiliary personnel, showing increases in the natural and social sciences as well as in the humanities between 1981 and 1985. Women represent 52 per cent in general, and in medical and social sciences more than 75 per cent of the auxiliary personnel in R & D institutions. However, in engineering and technology there is a marked decrease (-15 per cent between 1981 and 1985), which indicates that drawing is increasingly computerised.

At universities the share of women in leading functions has only recently reached the level of public debate. Female students' demands for women's studies topics to be taught by women professors were in some instances very spectacular and prompted such results as new appointments (e.g. a special chair in practical informatics at the Technical University of Vienna), while in other circumstances these demands caused delays in appointments and nominations or were not met at all.[4] In 1986, 18 per cent of university assistants were women, ranging from about 30 per cent in the humanities to 5 per cent in technical disciplines. Since 1980, 1,130 men but only 88 women have qualified as university readers ('Dozenten'), which means an average of 8 per cent per year.[5] Nevertheless the share of

---

4. After a long debate, a chair in 'Political science with particular reference to women's studies' was set up at the University of Innsbruck in 1988.

5. University teachers and researchers may be recruited in various ways. The career path generally is via the post of 'assistant' but only a certain proportion of assistants become fully qualified (with doctorate and a full teaching licence – the 'habilitation') for a permanent position at a university. The others drop out of the university system at various stages in the qualification process (about ten per cent every year). About ten per cent of assistants not in permanent posts are 'Dozent', i.e. they have a full teaching licence. 'Honorary professors' are scientists and specialists (mostly in the administration) who are awarded teaching authorisations by the collegial bodies of universities, thus gaining the rights of university readers, but having no compulsory teaching duties.

Table 1.3: Percentage of women scientists in universities and in other R & D-performing institutions, by fields of science and technology, 1985

Fields of science and technology	universities total (men and women)	%f	other R & D institutions total (men and women)	%f
Natural sciences	2 091	8.8	449	16.3
Engineering and technology	1 288	2.9	792	6.8
Medical sciences				
without hospitals	961	23.5	–	–
hospitals	1 765	18.6	–	–
with hospitals	2 726	20.3	171	29.2
Agricultural sciences	301	20.9	339	13.9
Subtotal				
without hospitals	4 641	11.0	–	–
with hospitals	6 406	13.1	1 751	12.6
Social sciences	1 303	21.0	453	21.6
Humanities	1 566	26.2	718	32.3
Subtotal	2 869	23.8	1 171	28.3
Total				
without hospitals	7 510	15.9	–	–
with hospitals	9 275	16.4	2 922	18.9

Source: Austrian Central Statistical Office, R & D Division

women who qualify as university readers has risen since the 1970s when it was about 3 per cent. In the academic year 1985/6 142 out of the 2,295 readers ('Dozenten') at Austrian universities were women (6 per cent). With professors, the percentage of women is still lower: 2 per cent of the full professors (19 out of 1,100), and 5 per cent of the senior readers (25 out of 516) are women.

An important influence in opening up access to positions of responsibility in the fields of science and technology is to be expected from the dominant group of senior researchers and directors of R & D units in general. And who was directing these R & D units at universities and other R & D-performing institutions in 1985?[6] Women constituted 7.5 per cent of the directors in the humanities (27 out of 361) and 5.8 per cent in the social sciences (i.e. 14 out of 241 research units in social sciences were directed by women). The corresponding proportions of women directors of research units were 3.3 per cent (9 out of 274) in the natural sciences, 1.5 per cent (3 out of 196) in the medical sciences and 1.4 per cent (4 out of 296) in the technical sciences. There was no female director in any of the 94 research units in the agricultural sciences. Apparently, in each field of science roughly one-third of the women scientists in the respective research units are directors of the unit. This figure corresponds to the proportion of manageresses in general employment or self-employment.

Employment trends for women scientists have not yet been assessed in the productive enterprise sector. The Federal Economic Chamber in Austria, having investigated R & D since 1966, in its ninth survey in accordance with the *Frascati Manual for Statistics on Scientific and Technological Activities* (Unesco R & D) began for the first time in 1987 to request gender-specific data also.

The results of the survey of the business enterprise (productive/enterprise R & D) sector indicate 11,216 men and 1,396 women engaged in R & D (see Table 1.4). In the report it is, however, emphasised that the 1,396 women researchers only represent a minimum. Apparently, the companies do not provide exact figures when asked to give a gender breakdown of employees. In general, however, there is a conspicuous trend towards higher qualified staff in R & D. While the total R & D staff in the productive/enterprise sector increased between 1984 and 1987 by 11.5 per cent, the increase was 13.9 per cent for scientists and 16.7 per cent for engineers.

Women's share in 1987 was 7.5 per cent out of 3,436 scientists (R & D full-time equivalent personnel) in the productive/enterprise sector, 7.1 per cent out of 6,304 technicians and 24.1 per cent out of 2,871 in the category of auxiliary personnel.

---

6. An important by-product of the R & D surveys of the Central Statistical Office is a comprehensive Directory of R & D Research Units in Austria. Latest edition: *Österreichischer Forschungsstättenkatalog 1986*, vols 1 and 2, eds Federal Ministry of Science and Research, Austrian Central Statistical Office, Federal Economic Chamber, Vienna, 1987, pp. 371 and 466.

Table 1.4: Percentage of women in enterprise R & D by type of work and qualification, by branch of activity (ISIC)[a], 1987 in full-time equivalent (FTE)[b]

ISIC Sectors/ selected branches	scientists/ engineers FTE	%f	technicians FTE	%f	auxiliary personnel FTE	%f	TOTAL FTE	%f
Mining	16.6	0	24.9	6.4	13.1	24.4	54.6	8.8
Manufacturing industries	3 413.5	7.5	6 265.2	7.0	2 844.8	24.0	12 523.5	11.0
Within selected branches:								
Electrical machinery	567.2	8.4	922.4	7.3	262.3	35.8	1 751.9	11.9
Electronic equipment & components	843.2	10.7	1 512.2	6.6	403.6	38.4	2 759.0	12.5
Chemicals	195.2	2.5	479.7	15.1	271.1	45.0	946.0	21.1
Drugs	113.1	14.5	220.4	21.0	139.8	44.9	473.3	26.5
Petroleum & refineries	28.3	2.5	46.4	0.2	21.1	0	95.8	0.8
Ships	20.0	0	22.4	0.5	21.0	18.1	63.4	6.2
Ferrous metal	156.7	1.0	178.8	1.3	147.2	3.2	482.7	1.8
Non-ferrous metal	17.0	2.4	45.1	7.3	23.6	10.2	85.7	7.1
Fabricated metal products	125.1	0.3	252.7	5.3	107.6	9.9	485.4	5.1
Instruments	10.4	0.4	27.5	7.3	5.3	43.4	43.2	10.9
Office & computing machinery	394.4	17.6	533.3	6.7	55.6	79.0	983.3	15.2
Machinery	327.5	1.3	727.6	2.0	426.5	5.9	1 481.6	3.0
Food, drink & tobacco	63.6	9.1	75.3	19.5	67.3	52.2	206.2	27.0
Textiles, footwear & leather	44.9	13.1	93.2	13.0	86.9	49.6	225.0	27.2
Rubber & plastic products	115.7	2.8	183.3	12.3	203.3	18.2	502.3	12.5
Stone, clay & glass	102.7	2.0	240.4	5.7	166.2	11.4	509.3	6.8
Paper & printing	46.2	0.9	84.5	17.5	87.4	18.8	218.1	14.5
Wood, cork & furniture	30.0	10	57.2	9.8	31.3	22.0	118.5	13.1
Services	22.7	4.9	39.1	10.5	26.7	16.9	88.5	11.0
Construction	20.3	5.4	38.0	10.8	26.7	16.9	85.0	11.4
Total	3 436.2	7.5	6 304.3	7.1	2 871.5	24.0	12 612.0	11.1

[a] ISIC stands for International Standard Industrial Classification of all Economic Activities.
[b] The FTE figure includes both men and women.
Source: Austrian Federal Economic Chamber.

Altogether somewhat more than 11 per cent out of 12,612 full-time equivalent personnel in Austria's productive enterprise R & D were women.[7]

There are, however, great differences in the proportion of economically active women according to the various branches of activity, particularly in the manufacturing industry. More women scientists and engineers (proportionately more than 10 per cent) are working, for example, in office & computing machinery, pharmaceuticals, textiles, wood and electronic equipment and components. The typical hierarchical pattern of the employment of women can be observed, for example, in the chemical industry. The proportion of women decreases with the type of work and level of qualification: 45 per cent of 271 auxiliary personnel, 15 per cent of 480 technicians and 2.5 per cent of 195 scientists. In the branch of office and computing machinery the employment situation looks promising for women graduates. In 1987, the proportion of women in this branch was 79 per cent of 55 auxiliary personnel, 6.7 per cent out of 533 technicians and 17.6 per cent out of 394 scientists.[8]

Concluding from this quantitative analysis of the various R & D-performing institutions and sectors in Austria, women scientists are, apparently in line with their general employment trends, on the move into science and technology. Most of them are economically active in their traditional fields of study, in the so-called soft sciences. However, some have opened the doors into the hard sciences, small in absolute number but gradually increasing. At universities, however, Austrian women have not yet reached the critical mass for decision-making or top positions in science and technology.

---

7. Summing up, the figures of both surveys (Austrian Central Statistical Office in 1985 and Austrian Federal Economic Chamber in 1984) indicated a total of 20,161 full-time equivalent personnel engaged in R & D, out of whom 26.5 per cent were in the higher education sector, 8.8 per cent in the government sector, 2.4 per cent in the private non-profit sector and 62.3 per cent in the business enterprise sector, which includes 6.2 per cent in each of the co-operative subsectors surveyed by the Central Statistical Office (i.e. co-operative research institutes, civil engineering & technical services as well as electricity-supply companies) and 56.1 per cent in the companies themselves (i.e. productive/ enterprise R & D sector) surveyed by the Federal Economic Chamber.

8. The number of personnel are calculated according to full-time equivalent (FTE).

Research Policy

The public debate on the promotion of research in Austria in 1967 prompted the Research Promotion Act, providing for two self-administrating research promotion funds[9] and, in 1970, the establishment of the Federal Ministry of Science and Research. The first woman Minister of Science and Research, Hertha Firnberg, determined science policy for thirteen years (1970-83) and also paved the way for the promotion of women.

The public promotion of research was embodied in law by the Research Organisation Act of 1981. This Act states the guiding principles for the promotion of research by the federal government as well as those of the organisation of federal research institutes. It established the Austrian Board and the Austrian Conference for Science and Research. Since 1982/3, under the presidency of the Minister of Science and Research, all of the thirteen Board members have been men. The members of the Conference (comprising scientists, politicians and the representatives of ministries, of provinces and of other interest groups in the economy, society and administration) are nominated from the various interest groups. Since 1982/3, out of forty-four members, four women represented administrations (e.g. Province of Vienna) and interest groups (e.g. Chamber of Labour); from 1986/7 onwards there have been six. Since 1981 the advisory bodies and expert groups on R & D matters of the two research promotion funds mentioned have also included workers' and employers' representatives.

The Rectors' Conference and the Fund for the Promotion of Research decide on research priorities, both in co-operation with each other and autonomously. While the Research Concept of 1972 was particularly concerned with building up research institutions, the subsequent Research Concept for the 1980s focused on the impact of research on socially relevant topics in the

---

9. The Fund for the Promotion of Research (Fonds zur Förderung der wissenschaftlichen Forschung, FWF) was established primarily to promote research which is not profit-oriented but which is aimed at the development of science in Austria. The second fund, the Fund for the Promotion of Research in Industry and Trade (Forschungsförderungsfonds für die gewerbliche Wirtschaft, FFF) was established for the promotion of research and development in industry and trade. The FWF is more important for university research. It fulfils its objective via the promotion of the research projects of individual research workers and research teams in different disciplines.

service of society, assuming the integration of science, research and technology policy. In the field of socially relevant research, a special priority research topic on 'women and research' was included.[10]

The access of women to positions of responsibility in careers of science and technology was pushed by nominations, designations and appointments. While all elected posts in the teaching and management of universities are dominated by men (i.e. twelve rectors, past rectors and rectors-elect as well as thirty-eight deans, with one exception during 1984/5), one out of the twelve university directors is a woman. (The head of the administration at a university is appointed by the Minister of Science and Research and is subordinate to her/him in all matters concerning public administration. The university director is also subordinate to the rector in matters concerning university autonomy.) Women in leading functions in individual divisions of the administration constitute 27 per cent, and in heads of deans' offices 65 per cent. Since 1983/4 the position of the Secretary General of the Rectors' Conference has been held by a woman.

With considerable time lag, however, the universities are on the move to eliminate the 'anti-feminine climate in, and masculine dominance of, academic circles' (Firnberg, 1987). Individual faculties at different universities assigned special working parties or women's delegates to promote the status of women in higher education as well as women's studies. The Rectors' Conference recently decided to set up and mandate a working party to consider possibilities of promoting female junior scien-

10. 'The scientific discussion on problems, experiences and interests of women should broaden and profoundly deepen knowledge and research. The scientific content of the disciplines should be examined as to whether and how they reflect the experiences and problems, activities and interests of women and, consequently, research approaches with regard to the content should be developed. In principle, all scientific disciplines should be involved, not only the social sciences and humanities, thus emphasizing important fields of future research work. Furthermore, research activities on women-specific problems should increasingly be carried out. For instance, in the following fields: economics; work; technology; state/politics; law; administration; medicine; information/mass media; education/socialisation; culture; history; religion; living conditions/day-to-day life; exclusion strategies. Women-specific questions should also be dealt with in research on social problems, e.g. in the areas of educational and occupational opportunities, working conditions, family, social security etc.' In: *Österreichische Forschungskonzeption 80*, p. 60 f.

tists and to make recommendations so as to dismantle the problems of access of qualified women to the scientific community. Discussions and decisions concerning temporary special measures have to take place at a lower level of university hierarchy so as to reach substantive numbers of women eligible for election.

In conclusion, nominations, appointments and elections of women scientists in the Austrian higher education system remain a vast field for further research and temporary special measures to accelerate de facto gender equity merit specific examination.[11] (Vogel-Polsky, 1989) It has to be recognised, however, that, following international recommendations, the public sector and the Federal Ministry of Science and Research have taken the lead.

Research Activities Related to Women

The public sector institutions were setting the pace for research activities related to the status of women. Research work on the education, training and careers of women and on women in the labour market started during the 1960s and has increased nationally and internationally particularly since the preparation of the International Women's Year 1975.[12] In Austria, this research was mainly commissioned by public authorities.[13]

Two comprehensive Government Reports on the Situation of

11. Cf. UN Convention on the Elimination of All Forms of Discrimination against Women: Article 4 states:
'1. Adoption by States Parties of temporary special measures aimed at accelerating de facto equality between men and women shall not be considered discrimination as defined in the present Convention, but shall in no way entail as a consequence the maintenance of unequal or separate standards; these measures shall be discontinued when the objectives of equality of opportunity and treatment have been achieved.
2. Adoption by States Parties of special measures, including those measures contained in the present Convention, aimed at protecting maternity shall not be considered discriminatory.'
12. Proclaimed in 1972 by United Nations Resolution A/34/3010.
13. By the Federal Ministries of Education, of Labour and Social Affairs, later on also by the Chamber of Labour and other interest groups, particularly in the fields of educational science, psychology, sociology, political science, subsequently in history, literature, law and economics. The findings provided the basis for numerous replies to the UN, UNESCO, ILO, OECD, Council of Europe, they were published in series edited by the mentioned Ministries or in journals of interest groups and stimulated action to promote the status of

Women in Austria (seven volumes each) were published by the Federal Chancellery in 1975 and 1985. Seventy-five per cent of the co-ordinators were women; the proportion of women in related research rose from 47 per cent in 1975 to 86 per cent in 1985.[14]

In September 1982, the Federal Ministry of Science and Research commissioned a comprehensive Research Documentation on Women in Austria on four topics:

1. Scientific activities on the theme 'woman' since 1970;
2. Investigation of all scientific work by women in the first half of the twentieth century;
3. A data bank on scientific activities, research findings and further publications on the theme 'woman' since 1975;
4. Documentation on individual women in sciences, arts and politics since 1900.

Today efforts are made from the onset to integrate the women's dimension in the design of research and women scientists in research teams, as the following example demonstrates. A comprehensive research priority topic on 'Handling the Crisis' was investigated by the Austrian Federal Ministry of Science and Research and involved Austrian social scientists of different disciplines and different institutional settings. Out of the original 26 individual projects, two were directed by women, and in projects dealing with personal data (person-

---

women and to pass legislation on gender equality. Cf. publications of the Division on labour relations and working women's affairs of the Austrian Federal Ministry of Labour and Social Affairs, mainly addressing, and accordingly distributed free of charge to, governmental bodies, workers' and employers' organisations, political parties, professional and women's organisations, libraries, media etc. As the Austrian Federal Ministry of Science and Research concluded in a paper on 'Women's studies at universities' (1986): 'University as an institution remained rather passive and aloof. The reasons for this detachment are not, perhaps, only or so much the topics of women's studies. Frequently, women's studies entail interdisciplinary and continuing education activities. Both, interdisciplinary work and continuing education still range at the fringe of universities esteem and endeavour.'

14. While the majority of researchers charged with the above-mentioned surveys and reports were increasingly women it seems interesting to note that in the 1960s and again in the mid-1980s (at the national and international level) some male social scientists focused their interest on women-related topics.

related) women and men researchers were often in parity. In the final publication the share of women comprised 13 out of 42 authors and 1 out of 5 members of the editorial team.

The results of an investigation as to whether the priority topic on 'women and research' was used to legitimate requests for funding indicate clearly that those institutions or scientific bodies which had previously initiated this topic and were accordingly informed about it, did refer to this point when increasing their funding. In particular, women scientists frequently referred to the priority topic on 'women and research' when applying for funding of research and/or teaching. The Federal Ministry of Science and Research will, evidently, also refer to this research topic and make the results available to the public either in the Ministry's library or by separate publications.

Otherwise, scientists or research institutions replied that women's themes came up with the demands of the women's movement and were accordingly investigated. Funding institutions indicated that, with the gradual increases in quantity, women's research also increased their qualitative aspects. For instance, the Austrian Fund for the Promotion of Research started to promote research on various aspects of women's studies in the early 1980s. These research projects were few in numbers and fragmented. However, in recent years the number of projects has risen, and a concentration on aspects of social history, social psychology, and theology was noted. In 1988 four groups of Austrian women researchers co-ordinated their work, and the Austrian Fund for the Promotion of Research supported four interrelated and interdisciplinary research projects on the role of women in the society of the nineteenth and the early twentieth centuries (history, social history, history of art, and literature).

When the Federal Ministry of Science and Research reports that the proportion of women applicants (who are usually project leaders) was 25 per cent in 1981/2 and decreased to 19 per cent in 1987/8, this does not, however, reflect the real position of women's share in projects. The majority of applications are forwarded from institutes directed by men while the actual research itself will be carried out by women. There is more scope for research into how and why the number of women scientists decreases in each category listed below:

applicants,
project workers,
project leaders,
expert referees (peer-review system),
members of executive boards deciding research policy in general or for the body on which they serve,
selection boards, etc.

The vast majority of the Austrian scientific community – not being involved in person-related research activities and not noticing the scarcity of women professors – belittled these developments on the fringes of academe, if they became aware of them at all, failing to react to the increasing number of women graduates as regards access to and promotion in R & D. In view of the heterogeneous and multidisciplinary objectives of research policy in general it seems helpful to negotiate national policy statements and corresponding research concepts in line with the international recommendations to promote the status of women.

Research Personnel Policy

Efforts to promote the status of women in the Austrian civil service should not be underestimated in R & D either. Implementing the recommendations of the International Women's Year, 1975 (United Nations World Action Plan, paragraph 62) a *Programme for Promotion of Women in the Federal Civil Service* was initiated in 1981 by the woman Secretary of State in the Federal Chancellery, Johanna Dohnal, and enacted by a decision of the Council of Ministers. This *Programme* provides that every minister appoints a working party for preparing a sectoral specific action plan and monitoring its implementation. With regard to recruitment policy, one of the objectives integrated into the action plans of all ministries was to selectively focus on increasing the share of women, especially in the civil service grades which are reserved for university graduates or grammar school graduates. In the individual ministries, the implementation of the *Programme* is mostly in the hands of so-called 'liaison or contact women'.

In 1982, a special working party was set up in the Austrian

Federal Ministry of Science and Research after consultation with the employees' representatives, comprising thirty-three members of the Ministry and fifty-six 'contact women' from sub-agencies (universities and higher education institutions, libraries, museums, scientific institutions, students' servicing authorities and students' counselling services). When designating these contact women, the emphasis was placed on including professors, assistants, scientific personnel and non-scientific personnel from all qualification levels. This working party meets five times a year, makes proposals for the advancement and training of women, organises information seminars for contact women, deals with personnel and legal questions, university legislation, etc. Subcommittees were set up, e.g. concerning the University Teaching Staff Act, further training, kindergarten, new technologies, working time, marketing for the realisation of the programme, etc. Between 1980 and 1985, women employees in the Federal Ministry of Science and Research increased by 25 per cent, compared with 12 per cent for men. In 1989 the contact women reported that they were particularly trying to motivate women to compete for leading positions.[15]

In conclusion, this *Programme for Promotion of Women in the Federal Civil Service* started in the field of R & D as a ridiculed initiative in 1981 and turned out to be a serious institution producing results in 1985.[16] In view of the projected output of qualified women from higher education and of their growing expectations to become economically active in their field of competence, the mentioned and other supportive measures seem to be necessary for quite a time so as to move the gates and gatekeepers with regard to the access to and promotion of qualified women scientists in R & D careers.

---

15. An interesting assessment is provided in this respect: Between 1 January 1986 and 31 December 1988, 12 leading positions (10 divisions, 2 groups) had vacancy openings. Out of 50 applicants 39 were men and 11 women. However, 11 of the leading posts were finally occupied by a man and one by a woman.

16. The Working Party of the Austrian Federal Ministry of Science and Research is presided over by Director Dr Edith Fischer (Head of Division of scientific libraries, documentation and information, including Unesco). This *Programme for Promotion of Women in the Federal Civil Service* was initially to be terminated in 1985, but was prolonged by decision of the Cabinet Meeting for an indefinite period.

Dual Career Couples in R & D

Since the 1950s gender roles and role expectations of women and men, their perceptions of themselves and of the opposite sex, their attitudes and behaviour have been repeatedly investigated. Every new generation of women getting involved in women's movements started again from 'square one' during the following decade. Professional women, be they in the fields of higher education, engineering and technology, medicine or literature and the arts, are obviously a motivated select group who manage to reject stereotypes of men's and women's jobs, on the one hand, and to reconcile professional and family roles, on the other. (Gaudart, 1975)

In general, however, not much research addresses the complementary behaviour patterns of professional men, be they relatives (fathers, spouses, sons) and/or colleagues. It was only during the 1970s that the relative position of women and men was investigated, and in about 1975 the question of representation of interests and participation of women in decision-making processes was put on the research agenda. (Gaudart and Greve, 1980)

It might be assumed that the situation of scientific workers and their working conditions do not differ so much from that of other professional work, in particular as regards the gender perspective. Obviously, the science sector becomes increasingly an institutionalised and regulated social fabric. The production of knowledge, the interaction in scientific institutions, the different forms of co-operation have been continuously investigated (cf. Knorr-Cetina, 1981; Dickinson, 1986) recently including also the ethical debate (de Hemptinne, 1989). Investigations rarely distinguish between men and women. (Stolte-Heiskanen, 1983) Research interests have focused on the genderless individuals' contribution to, and participation in, scientific progress. The question was not turned around so as to inquire into the impact of scientific work on the personal life of a scientist, and even less so whether different approaches would be necessary as to the gender of the scientist.

Based on studies of scientific careers by Eiduson (1962) and Cattell (1963), List (1985) refers to their assumptions that the obvious and rather trivial difference between professional and scientific workers were the latter's inclination and ability to do scientific work. The personality characteristics of a scientist were mainly related to the role expectation of a scientist in the societal context. Three sets of characteristics were distinguished: (1) intelligence and radicalism *vis à vis* opinion authorities, i.e. an intellectual habit of scepticism and criticism, which means intrinsic cognitive requirements; (2) dominance and independence, personal traits usually associated with professional men in leading positions; and (3) introversion and self-control. Undeniably these characteristics can also be expected from women scientists. However, if women scientists practise more dominant behaviour they should be aware of rising problems in changing gender relations.

According to behavioural psychologists (Morley, 1980) these demands on women scientists for their integration into the men's world create stress; this is also true for the dominant scientists who grew up in a culture where the subordination of women was part of the 'natural' order of things – the way things 'should be'. The dominant scientists, i.e. men, learned a set of traditional attitudes and expectations about their place in the scientific world, both in relation to women and to colleagues. However, since the 1960s, assumptions about the dominant place of men in society, and about the structure of their scientific world, have changed substantially. The dominant scientists are told that the inferior position of their women colleagues is no longer an 'acknowledged' fact, but a major flaw in social structures which must be changed. Because these dominant scientists yield so much power on all levels of the scientific communities, a change in this organisational structure is being called for. This introduces a new dimension of responsibility and complexity into their scientific and working lives. Future research projects should thus not be directed again towards gender differences but towards work organisation in the scientific community and possible alternatives.

Scientific interests develop early. It was proved for many professions that there is a kind of heritage of professions from the parent which continues in choosing partners. Not only the disciplines but also the levels of education were the same for the professional women and their fathers or spouses. (Gaudart, 1975; Diem-Wille, 1989)

Accordingly, the scientific community might be confronted with another emerging trend. Since more women are qualified and motivated to be fully active in their fields of competence, they increasingly find their spouses in the professional community where they work. Austrian microcensus and other survey data on social contacts in metropolitan areas indicate that 30 per cent of all people find their spouses – while 30 per cent find a new partner – at their workplace. (Schulz, 1978) The phenomenon of the dual career couple, where both partners share a deep commitment to simultaneously maintaining their personal relationship and their careers, has been increasingly documented in the 1980s. (Hootsmans, 1987)

Mesdames Marie Curie and Irène Joliot-Curie were working with their husbands. Such scientifically productive husband-and-wife teams are early illustrations of this kind of scientific tandem careers. Besides, the mother provided a role model for the daughter. Dual career couples among R & D personnel could be an interesting subject not only for the female/male scientists involved but also for their employers, whether in higher education, government or non-profit sectors. The scientific community might ask itself how it copes with these developments and how it plans its personnel development programmes.

Conclusions

Women are on the move in science and technology and the demand for research is expected to grow. The overall objective being the integration and participation of women, women scientists should become instrumental in the general process of science and not merely struggle for personal emancipation by research related to women. Major problems in working conditions as well as in the responsibilities and ethics of scientific workers lie ahead on the way towards the turn of the millenium.

In order to initiate a major effort at different levels it seems necessary, firstly, to develop a research policy in favour of women scientists, and secondly, to investigate the scientific community on the basis of this objective.

The impact of science policy on women should be monitored and evaluated periodically, on the national level, e.g. by questions to Members of Parliament and, on the international level, UNESCO should be invited to continue examining these objectives, however, not only separately but also by integrating them into each of its major fields of activity and in all affiliated scientific societies, councils, networks and the like.

References

Austrian Central Statistical Office (ed.), 'Census 1981: Educational level of the population', Messmann, K., in *Statistiche Nachrichten*, Heft 12/1986.

Austrian Federal Economic Chamber, Division for Statistics and Documentation (ed.), *Forschung und Dokumentation in Österreich 1987* (Research and Documentation in Austria). Vienna, 1989.

Austrian Federal Ministry of Science and Research, Central Statistical Office, Federal Economic Chamber (eds), *Österreichischer Forschungsstättenkatalog 1986* (Austrian Research Institutes Directory). Vienna, 1987, vols 1 and 2, pp. 371 and 466.

Austrian Federal Ministry of Science and Research, *Bericht des Bundesministeriums für Wissenschaft und Forschung über die Durchführung des Förderungsprogrammes für Frauen im Bundesdienst* (Report on the Implementation of the Promotional Programme for Women in Civil Service), Vienna, 1989.

——(ed.), *Frau und Universität* (Women and the University), Sonderdruck aus Hochschulbericht, Knollmayer, E., Vienna, 1981.

——,*Österreichische Forschungskonzeption 80* (Austrian Research Plan). Vienna, n.d.

——,'Women's studies at universities', paper prepared by E. Hack for European Workshop on Post-Graduate Studies and Women's Studies. Lisse, the Netherlands, Nov. 1986.

BURGER, R. et al., *Verarbeitungsmechanismen der Krise* (Handling the Crisis). Vienna: Braumüller, 1988.

CATTELL, R.B., 'The Personality and Motivation of the Researcher from Measurement and from Biography', 1963.

DE HEMPTINNE, Y., 'Ethical constraints and their integration into national research policies: A review of present thinking', in *Progrès scientifique et débat éthique*, Plaidoyer pour l'analyse politique, sous la direction de Bernard Crousse et Luc Rouban. Paris: Editions du Cerf, 1989.

DICKINSON, J.P. (ed.), *Science and scientific researchers in modern society.* Paris: Unesco, 1986, p. 260.

DIEM-WILLE, G., 'Karrierefrauen und Karrieremänner' (Career women and career men), A psychoanalytical analysis of biographies and family dynamics, Preliminary Report, Vienna, 1989.

EIDUSON, B.T., *Scientists: Their Psychological World*, New York, 1962
FIRNBERG, H. AND L.S. RUTSCHKA, *Die Frau in Österreich* (Women in Austria), Vienna, Austrian Federation of Trade Unions, 1967, p. 119.
FIRNBERG, H., 'Frauen und Forschung' (Women and Research) in *Frauenstudium und akademische Frauenarbeit in Österreich (Women's higher education and academic work in Austria) 1968–1987*. Vienna, Austrian Federation of University Women, 1967, pp. 17–29.
GAUDART, D., *Zugang von Mädchen und Frauen zu technischen Berufen* (Access of Girls and Women to Technical Careers), in A. Niegl (ed), Series on the Education of Girls and Women vol. 3. Vienna: Östereichischer Bundesverlag, 1975, p. 527.
GAUDART, D. AND R. GREVE, *Women and Industrial Relations*, Framework paper and analysis of the discussions of an international symposium (Vienna, September 1978), Research Series, No. 54. Geneva: International Institute for Labour Studies, 1980.
GROSS, I., In *Disparities of living conditions among women and men in Austria*, Statistical analysis. Vienna: Austrian Federal Ministry of Labour and Social Affairs, 1989.
HOOTSMANS, H., 'Two Career Partnerships: Having it all or balancing priorities', in *The Interface of Work and Family: Two Career Couples Rewrite the Rules*, Panel, at the Third International Interdisciplinary Congress of Women – Women's Worlds: Visions and Revisions. Dublin, Trinity College, July 6–10, 1987.
KNORR-CETINA, K.D., *The Manufacture of Knowledge, An Essay on the Constructivist and Contextual Nature of Science*. Oxford: Pergamon, 1981.
LIST, E., 'Der asketische Eros, Über Geschichte und Struktur des wissenschaftlichen Habitus' (The ascetic eros, on the history and structure of the scientist's habitus), in K. Freisitzer *et al.* (eds), *Tradition und Herausforderung. 400 Jahre Universität Graz* (Tradition and Challenges. 400 Years University of Graz). Graz, 1985, pp. 407–22.
MAURER, M., *Feministische Kritik an Naturwissenschaft und Technik. Eine Einführung* (Feminist criticism of Natural and Technical Sciences. An Introduction). Hochschuldidaktische Arbeitspapiere 23, Hamburg: Interdisziplinäres Zentrum für Hochschuldidaktik der Universität Hamburg, 1989.
MORLEY, E., 'Men's Careers, Women's Careers, and the Need for Intelligent Change Strategies', in I. Lamel, *Management Careers in Changing Societies*. Vienna: Austrian Federal Ministry of Social Affairs, 12, 1980, pp. 51–80.
OECD, *Reviews of National Science and Technology Policy, Austria*. Paris: OECD, 1988.
SCHULZ, W., *Sozialkontakte in der Großstadt* (Social contacts in metropolitan areas). Vienna: Institut für Stadtforschung, 1978.

STOLTE-HEISKANEN, V., 'The Role and Status of Women Scientific Research Workers in Research Groups', in J.H Pleck & H. Lopata (eds), *Research in the Interweave of Social Roles*, 3, 1983, pp. 47–56.

SUTHERLAND, M.B., 'The Role of Women in Higher Education', *Higher Education in Europe*, vol. 10, no. 3, 1985, pp. 46–85.

Unesco/CEPES: *Higher Education in Austria*. First published as a monograph on higher education by Unesco/CEPES 1987. Vienna: Federal Ministry of Science and Research, 1988.

Unesco: *Manual for Statistics on Scientific and Technological Activities*, ST-84/Ws/12. Paris, 1984.

VOGEL-POLSKY, E., *Les actions positives et les contraintes constitutionnelles et législatives qui pèsent sur leur mise en œuvre dans les Etats membres du Conseil de l'Europe*. Strasbourg: Conseil de l'Europe, EG(89)1.

2

Handmaidens of the 'Knowledge Class'.
Women in Science in Finland[1]

Veronica Stolte-Heiskanen

It is difficult to say who holds the power in a precise sense, but it is easy to see who lacks power.
  M. Foucault, Language, Counter-Memory, and Practice

Introduction

Advanced industrialised societies have experienced a dramatic expansion of education, especially higher education, and large-scale transformations in the occupational structure during this century. With the steadily growing share of the population engaged in 'mental work' and production of goods and services, the educated middle class with specialised formal education has grown to a sizeable minority. Because of the growing importance of educated labour in the production of knowledge, in the control functions of organisations and administration and in the maintenance and reproduction of the population, post-industrial society is often described by the rise of the new class (Gouldner, 1979) or the 'knowledge class' (Westoby, 1987).

In Finland, too, the proportion of the over fifteen-year-old population who graduated from second or third level educational institutions had grown from 13 per cent in 1960 to 46 per cent by 1985, when 28 per cent of the population was employed in the so-called knowledge class occupations.[2] The proportion of women in these occupations increased at least as rapidly and

---

1. Data for this study were collected with the valuable assistance of Marja Särkilahti.
2. In official statistics defined as the category of *Community, social and personal*

more recently even faster than that of men. In 1980 more than one third of all economically active women were in these occupations and in 1985, women constituted more than two thirds of the economically active in this sector. (Position of Women, 1984; Population Census, 1988)

As Gouldner (1979) notes, in contemporary society the certification to have competent authority occurs through education. Thus the educated professionals are increasingly cornering the markets in expertise. This may be particularly true of the science and technology professions, since in post-industrial societies productivity more and more depends on science and technology. There is also a widespread belief that society's problems are solvable on a technological basis and with the use of educationally acquired technological competence. Because of the growing importance of science and technology in society, scientists and engineers constitute an increasingly significant subgroup of the knowledge class.

Disco (1987) identifies three crucial demarcation criteria of the knowledge class: (1) performance criteria, based on individual achievements, (2) inheritance criteria, such as race, ethnicity, family background, culture, and (3) 'chance' criteria, such as gender, and other physical distinctions or special capacities. At least in science, there is strong evidence that the chance criterion of gender is universally applied to place women on the periphery of science. In Finland it certainly takes precedence over inheritance criteria.

This study examines to what extent the chance criterion of gender plays a role in the position of women in Finnish science.

The Road to a Scientific Career:
Women and Higher Education

Up until the mid-1960s there was a steady increase in the number of new enrolments at institutions of higher education because of the growing share of the population passing the matriculation examination (ISCED level 3).[3] Consequently, the

---
services, including public administration, education, research, health and professional personal services.

3. Here and in the following pages the educational levels correspond to the definition given in Unesco: *International Standard Classification of Education*.

higher education system was rapidly expanded. By the beginning of the 1970s a network of seventeen higher educational institutions was created. Today there are ten universities with several faculties, three universities of technology, three schools of economics and business administration, and a college of veterinary medicine.

For more than two decades the proportion of women university students has been somewhat over 50 per cent. By 1986, more than half of the university degrees (ISCED level 6) were granted to women and their share among post-graduate students rose to 41 per cent in 1986. The proportion of women obtaining a post-graduate degree has more than tripled in the last two decades: in 1986, 30 per cent of the post-graduate degree recipients were women.[4] (Korkeakoulut, 1988)

Despite the exceptionally high rate of women's participation in higher education, their distribution according to field of studies does not significantly differ from other Western European countries. The universally noted phenomenon of gender differences in the field of studies is also evident in Finland.

As a result of conscious attempts to redistribute the flow of students to different fields in accordance with the new demands for scientific manpower, by 1980 there was a substantial increase in the number of students in the natural, technical and medical sciences and a corresponding decrease in the social sciences and humanities. This, however, did not significantly affect the distribution of women in different fields. The proportion of women graduates grew most in the already female-dominated fields. The proportion of women in different fields of studies is shown in Table 2.1.[5]

---

4. In Finland there are two post-graduate degrees. Both the first (licentiate) and second (Ph.D.) level require independent research and dissertation. The latter dissertation must be published and publicly defended.

5. The high proportion of women in natural sciences is misleading, since their share in different sub-disciplines greatly varies. The interest of women in different subfields of natural sciences follows the internationally observed sequence: biology, chemistry and biochemistry, and the share of women in these subfields has not changed much during the last decade. Between 1975 and 1984 in the traditionally female-dominated field of pharmacy and, to a certain extent, biology, the proportion of women has increased, while in the male-dominated fields of computer sciences and physics it has even slightly decreased (Koskiala, 1986). Similarly, within the engineering faculties women are more concentrated in architecture than engineering and technology: in 1986 the proportion of women in the former was 40 % and the latter 14 % (Korkeakoulut, 1988).

Table 2.1: Women's participation in higher education, by field of studies 1985/6

Field	New students	women MA degrees awarded[a]	Post-graduate degrees awarded
Humanities	73	71	48
Social sciences	60	56	35
Agricultural sciences	50	52	32
Medical sciences	73	70	34
Natural sciences	40	49	26
Engineering and technology	19	15	14
Total number	(1 3078)	(8 165)	(681)
% women	53	52	30

Source: Korkeakoulut, 1988.
a. Includes 1 627 lower level undergraduate degrees.

The scarcity of women among natural and technical science students seems to be more the result of self-selection than conscious discriminatory practices of the higher education system. There is hardly any difference in the percentage of women participating in the entrance examinations and that of those accepted for enrolment to the different faculties. (Komiteanmietintö, 1988)

The Position of Women on the Scientific Market

There are no systematic data available on the number of women scientists and engineers actually engaged in R & D. It is, however, generally well known that, unlike in other knowledge class occupations, their share is rather small. In the traditional institutional hierarchy of the Finnish scientific community the universities represent the centre of the scholarly field. Compared to other scientific institutions they have shown the most liberal attitudes towards the presence of women academics – although women are far from equal even in academe.

Earlier studies have shown that women are more tolerated in rapidly expanding fields. (Rossiter, 1978) This also seems to be true in the case of institutional expansion. In the course of the rapid expansion of the higher education system in the 1960s the number of professorships more than tripled. There was also a substantial increase in the number of middle and lower level academic posts. During the expansive phase women had the best opportunities to fill open vacancies: the proportion of women among newly appointed professors reached an all-time high of 12 per cent in 1979, and among associate professors 16 per cent in 1982. (Seurantatyöryhmän mietintö, 1986) When the growth of the universities stopped after the mid-1970s, increased competition and greater scarcity of jobs reduced especially women's hopes for an academic career. The number of women obtaining doctorates almost doubled during the 1980s, while the corresponding increase among men was only 29 per cent. (Luukkonen-Gronow, 1987) Yet in 1988, women still constituted only 7 per cent of the professors and 11 per cent of the associate professors. (Suomen Valtiokalenteri, 1989)

For the past two decades the greatest increases in academic posts have been in the technical, natural, medical and social sciences. (Räty and Luukkonen-Gronow, 1981) The increase in the number of posts in these fields has not been accompanied by a corresponding growth in the share of women. The number of women increased most in those fields where they were already the most numerous. The distribution of academic women scientists by field and position is shown in Figure 2.1.

In addition to the universities, important and much coveted research posts are the research fellowships of the Academy of Finland.[6] As a rule, Academy researchers are situated at the universities and are otherwise, too, integrated in the academic community. As can be seen from Table 2.2, the position of women in Academy research posts is similar to that of the universities.

In every post category the proportion of women increased up until the beginning of the 1980s, after which it became stabilised around the 1983 level. Again, the higher one goes up the hierarchy, the fewer the women. However, only in the case of research assistants, where the competition for posts is greatest, is the proportion of women appointed relatively smaller than

---

6. The structure of the academic teaching posts and corresponding research fellowships of the Academy of Finland (SA) is illustrated in the chapter Appendix.

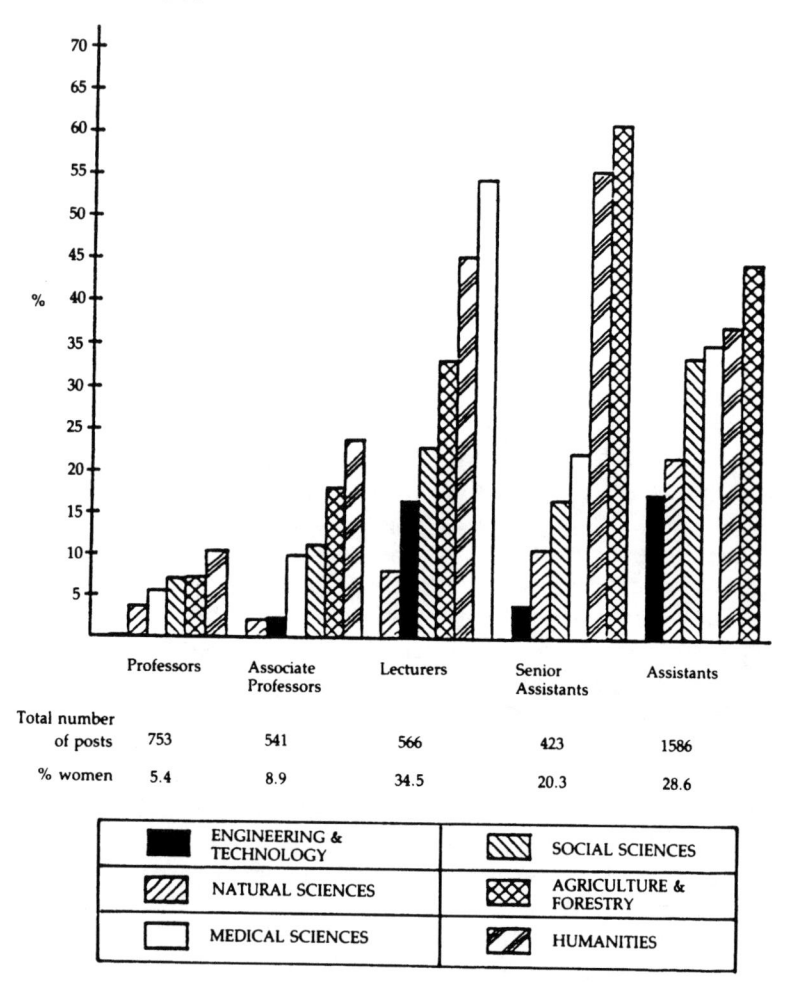

Figure 2.1: Distribution of academic women by field and position in 1985

	Professors	Associate Professors	Lecturers	Senior Assistants	Assistants
Total number of posts	753	541	566	423	1586
% women	5.4	8.9	34.5	20.3	28.6

- ENGINEERING & TECHNOLOGY
- NATURAL SCIENCES
- MEDICAL SCIENCES
- SOCIAL SCIENCES
- AGRICULTURE & FORESTRY
- HUMANITIES

Source: Seurantatyöryhmän mietintö, 1986.

their share of applicants. (Komiteanmietintö, 1982; Seurantatyöryhmän mientintö, 1986) Although women are somewhat more evenly distributed among different fields, the familiar concentration into 'female specialties' is evident here, too. The percentage of women in all research posts is smallest in engineering and technical sciences and in the social and natural sciences.

Table 2.2: The share of women in research posts of the Academy of Finland

Post category	1966 %F	1966 No. Sc.[a]	1973 %F	1973 No. Sc.	1980 %F	1980 No. Sc.	1983 %F	1983 No. Sc.	1985 %F	1985 No. Sc.	1987 %F	1987 No. Sc.
Senior researchers	9	(20)	10	(40)	20	(46)	17	(77)	19	(84)	21	(88)
Junior researchers	17	(42)	14	(79)	18	(85)	27	(111)	28	(123)	26	(131)
Research assistants	11	(71)	20	(163)	23	(202)	36	(207)	34	(226)	34	(224)
Total	13	(133)	17	(282)	21	(333)	30	(395)	29	(433)	29	(443)

Source: Seurantatyöryhmän mietintö, 1986.
a. No. Sc. = Number of Scientists

Table 2.3: Percentage of women scientists at state research institutes[a]

Position	1980 %F	No. Sc.[b]	1985 %F	No. Sc.	1988 %F	No. Sc.
Directors	0	15	0	15	0	16
Department heads	4	84	5	98	14	158
Researchers	18	988	22	1 346	20	1 571

Sources: For 1980 and 1985 data from *Seurantatyöryhmän mietintö*, 1986, for 1988 *Suomen Valtiokalenteri* 1989 and Technical Research Center of Finland.
a. Included are only those state research institutes where more than 30% of their annual budget is used for R & D. On the basis of this criterion two new research institutes were added and one formerly included was left out of the 1988 figures.
b. No. Sc. = Number of Scientists

While the position of women at the state research institutes has not been extensively studied, opportunities for women's research careers seem to be even more restricted there. There is evidence of increasing gender demarcation at each step of the hierarchy here too.

Compared to the universities, on each level of the hierarchy there are fewer women. There has been no significant improvement in the proportion of women throughout the 1980s.

No systematic data exist on the private sector, but there is some evidence that the number and position of women scientists and engineers is even lower there. Studies of university graduate professionals (Saari and Majander, 1985; Silius, 1989), a recent survey of the members of the League of Engineers and Architects (Hassi, 1986), as well as a study of research groups working in different institutional settings (Stolte-Heiskanen, 1981), all suggest that women scientists and professionals tend to be concentrated in the public sector. In the absence of data it is impossible to say whether this is a result of conscious self-selection or gender discrimination, or both.

Women in science are not only numerically under-represented but they also experience greater difficulties in embarking on a scientific career. Earlier studies, for example, in the USA, have shown that women scientists have more difficulty finding employment, have fewer possibilities of promotion and access to fewer supervisory posts. (Zuckerman and Cole, 1975; Vetter *et al.*, 1978) This also seems to be the case in Finland. More than half of the women, compared to almost one-quarter of the male engineers and architects, had difficulties in getting a permanent job corresponding to their educational qualifications after graduation. (Hassi, 1986) A retrospective study of the career patterns of male and female 'academic influentials' also found that women experienced greater problems in getting a permanent job at the beginning of their career. Men frequently advanced by radical jumps up the ladder of the hierarchy, while women's mobility occurred through conventional step by step advancement. (Häyrynen, 1988) As such, it is not surprising that women scientists have a lower feeling of job security and more frequently anticipate difficulties in finding similar or better positions should they leave their present jobs. (Stolte-Heiskanen, 1983)

Beyond the 1987 Law of Equal Rights, which also applies to the science and technology professions, there are no specific science policy measures to promote or guarantee gender equalities in science. In 1981 the Ministry of Education appointed a subcommittee for investigating the problems in women's research careers. The subcommittee's recommendations included a variety of measures to promote the increased participation of women in post-graduate training, greater equality in appointments to research posts and the provision of adequate social services (such as day care centres) to enable women to effectively pursue research careers. (Komiteanmietintö, 1982) Although these recommendations received wide public attention, as the follow-up subcommittee's report concludes, 'there is no evidence that the position of women as scientists has improved during the past three years'. (Seurantatyöryhmän mietintö, 1986:7)

Why So Few Women?

One of the frequently advanced explanations for the relative absence of women from science is that socialisation processes reproducing traditional gender roles result in the self-selective avoidance of scientific careers by women. It has been suggested that the masculine image of science leads to women's weak scientific orientation, on the one hand, and to a lack of self-confidence in their scientific potential, on the other. (Zuckerman and Cole, 1975; Vetter et al., 1978) Among Finnish university students, women indeed were found to be less certain than men of their career choices, and they became even more uncertain as they advanced in their studies. While both men and women consider, next to security, creativity as the most important factor in career selection, women rate scientific abilities as being of considerably less importance than men do. (Häyrynen, 1984)

Also professional women scientists tend to evaluate their personal qualifications less highly than men (Saari and Majander, 1985) and rate their theoretical abilities lower than their male colleagues (Stolte-Heiskanen, 1982; Häyrynen, 1984). Women themselves are more often of the opinion than their male colleagues that 'women's lower status in science is due to their lack of self-confidence'. (Stolte-Heiskanen, 1982; Julkunen, 1982; Bergman, 1985)

Perhaps the most firmly held universal belief is that the demands of a scientific career are incompatible with the socially valued roles of wife and mother. The role incompatibility thesis does not find support from the Finnish data. Unlike in many Western countries, in Finland the majority of women scientists are married, although a slightly higher proportion of women than men are single. (Haavio-Mannila, 1981; Julkunen, 1982; Luukkonen-Gronow and Stolte-Heiskanen, 1983; Bergman, 1985) Almost as many women as men are married and the majority have children at the time they receive their doctorate. (Luukkonen-Gronow, 1987)

This is not to say that marriage and a family does not constitute a greater liability for women than for men. The greatest pressure for achievement and embarking on a scientific career coincides with the establishment of a home and the family: between the ages of twenty-five and thirty-five. This probably explains why women are older than men at each step of the conventional career pattern.

Women scholars with young children also face problems of keeping long and irregular working hours. Women scientists have been systematically found to work fewer irregular hours than men, and the number of hours worked by women decreases with every child they have. Maternity leave may often be disruptive to research in progress. Consequently, some women prefer not to take advantage of the full maternity leave benefits and return to work earlier.[7]

In the households of most women scientists too, the traditional division of labour prevails, and women primarily bear the burden of household and childrearing tasks. Thus, unlike their male colleagues, women have to cope with a double work load. On the other hand, for many women the family constitutes an important resource. They often receive compensatory support from the spouse and family life as such for the disadvantages and emotional stress experienced in their professional environment. (Luukkonen-Gronow, 1984; 1987)

'Old Boys' and Handmaidens: Experiences in the Scientific Community

Qualitative differences in scientific activities are often, but not always, associated with different positions in the scientific hierarchy. Some scientific activities are more prestigious and bring more rewards than others. While differences in the tasks performed by scientists are partly accountable by their different hierarchical positions and differences in qualifications, there is strong evidence that gender often plays a role in a selective process which assigns women to less rewarding tasks.

In Saari and Majander's study (1985) the largest percentage of men were working in planning and development, while almost a quarter of the employed women did primarily clerical work. Similarly, about one-third of the male architects and engineers but less than a quarter of the females worked in materially more lucrative administrative and commercial jobs, while twice as many women as men were engaged in relatively routine laboratory R & D work. (Hassi, 1986) Among scientists holding formally equal positions in research groups, men generally indicated higher involvement in creative tasks (e.g. identification of research problems, formulation of hypotheses, methods, and interpretation of results), while women had higher involvement in routine activities, such as routine analyses, data collection and literature review. Men also tend to

---

7. Women are entitled to approximately one year's maternity leave. In principle temporary academic appointments are extendable to a period corresponding to the length of the maternity leave.

participate more in the planning of research in their unit.[8] (Stolte-Heiskanen, 1983) In another study based on solicited career histories of women scientists, women reported that they are more often given secretarial tasks than their male counterparts, and consequently they are often considered 'fit to make coffee and settle conflicts', but their research does not excite interest.[9] (Luukkonen-Gronow, 1984)

Likewise, women's professional roles tend to be more narrowly confined to research tasks within their immediate work environment. Considerably fewer women are entrusted with administrative tasks or are called upon to act in expert or other professional roles, than their male colleagues. Women in, for example, research groups participate less often in research projects outside their own unit, take part less frequently in national scientific conferences, seminars, etc., or are less likely active members in scientific associations. (Stolte-Heiskanen, 1982; 1983; Bergman, 1985)

Membership in what Cole (1979) has called the 'old boys' network' is not only a significant source of scientific capital but, by sharing socially important information such as regarding grants, vacancies, ongoing professional activities, etc., it also contributes to the acquisition of social capital. (Morgan, 1981; Bourdieu, 1988) The existence of homosocial mechanisms reproducing the dominant male culture of the scientific community noted in previous studies (Lipman-Blumen, 1976; Reskin, 1978) is also evident in Finland. In a sample of young scientists, the majority of both men and women frequently discuss work-related topics with colleagues of the same sex. However, almost half of the women but less than a third of the men have frequent work-related discussions with colleagues of the opposite sex. (Stolte-Heiskanen, 1982) Married men especially seem to limit their professional discussions to members of their own sex. Similar results were obtained in Bergman's study (1985).

8. Similar results were obtained in five other countries participating in this comparative study: Austria, Belgium, Hungary, Poland and Sweden. Men and women scientists were more or less of the same age and educational background and had been about the same length of time in the unit (see Stolte-Heiskanen, 1983).

9. The study is based on letters describing the career histories of women scientists solicited by the Subcommittee on Problems of Women's Scientific Careers, to which eighty-three women scientists responded (see Luukkonen-Gronow, 1984).

Data from the career histories of women also reveal condescending attitudes towards women's scientific abilities by their male colleagues. Over 20 per cent of the respondents encountered attitudes of not taking seriously the research done by women. Unmarried women, in particular, experienced these types of depreciating attitudes. (Luukkonen-Gronow, 1987) Also many female engineers felt that their work was not appreciated or rewarded by the same criteria as those of their male colleagues. Their work often remained invisible and was sometimes even presented under someone else's name. In Hassi's study (1986) not one of the respondents felt that they were considered equal to their male colleagues. Similar experiences have been reported for British women scientists (Ferry and Moore, 1981) and US women engineers (Kirkham and Thompson, 1984). In another study of women engineers, almost half of the respondents have even experienced at some time or another at least mild forms of sexual harassment from their male colleagues. (Haavio-Mannila, 1987)

The Scientific Power Structure: Where are the Women?

Perceived and real differences in the roles assigned to women in science easily lead to their perception by their male colleagues in terms of stereotyped categories and to the lack of recognition of their work. Their competence and achievements remain invisible outside the subdominant culture of women scientists. For example, in a study of acknowledgement networks in doctoral dissertation prefaces, women were found to acknowledge women scientists' contributions twice as often as men. For the latter women remain invisible. (Bruun *et al.*, 1982)

Invisibility leads to lack of scientific credibility, which in turn limits access to recognition. Within the reward system of science one traditional indication of the recognition given to scientists is to be elected or invited to positions of influence and leadership in the institutional infrastructures of the scientific community. Table 2.4 summarises to what extent women have access to capital of scientific power and prestige. (Bourdieu, 1988)

In principle officers of national scientific associations are elected by members on the basis of their recognised merits and cognitive authority. Only 11 per cent of the 125 scientific associations

Table 2.4: Percentage of women in positions of scientific power and prestige

Organisation and position	% Women in organisation or position	% women in sub-fields	Number of persons (or units)
Scientific associations (1988)[a]			(125)
Chairperson	11		
Secretary	35		
Board member (1985)	20		(122)
Humanities		24	301
Social sciences		21	163
Natural sciences		19	247
Agricultural sciences		21	39
Engineering and technology		12	195
Medical sciences		30	90
Scientific journals (1985)[b]			
Editor or editorial board member	19		(83)
Humanities		17	81
Social sciences		21	84
Natural sciences		20	157
Agricultural sciences		15	39
Engineering and technology		17	87
Medical sciences		16	50
Finnish Academy of Sciences (1958–1988)[c]			
Distinguished scientist award recipient (1958–83)	0		49
Member	7		344
Natural sciences		4	197
Humanities		12	147
Foreign members (1958–83)	0		185
Scientific Society of Finland (1988)[d]			
Member	7		180
Exact sciences		2	48
Life sciences		4	47
Humanities		8	47
Social sciences		16	38
Honorary member	0		8

Table 2.4 continued

Organisation and position	% Women in organisation or position	% women in sub-fields	Number of persons (or units)
Expert referee in professorial appointments in selected fields (1950–1985)[e]	12		(99)
Education sciences		35	(31)
History		3	(30)
Business economics		0	(26)
Sociology		0	(12)

Source: Antikainen and Jolkkonen, 1988.
a Units refer to the number of associations. Calculated from Suomen Valtiokalenteri 1989. Data on board members from Seurantatyöryhmän mietintö, 1986.
b Source: Seurantatyöryhmän mietintö, 1986. Units refer to number of journals.
c Data for 1958–83 calculated from Halila, 1987, for 1988 Suomen Valtiokalenteri, 1989.
d Source: Suomen Valtiokalenteri 1989.
e Units refer to the number of professorial appointments.

surveyed entrust leadership to a woman and, as the distribution of board members shows, the proportion of women is low even in the more female-dominated fields. The situation is almost identical in the case of the eighty-three national scientific journals published in a variety of disciplines.

The Finnish Academy of Science and its equivalent for Swedish-speaking scholars, the Scientific Society of Finland, are modelled on the old European tradition of 'academies' of distinguished scholars. Life membership is by invitation, which signifies formal recognition of a scientist's achievements. In the case of both scientific societies, women constitute only 7 per cent of all living members. For a quarter of a century the Finnish Academy of Science has not seen fit to invite a single woman foreign member or to honour a woman with the annual distinguished scientist's award. The first woman to gain admittance to these 'old boys' clubs came from the humanities in 1949 (Finnish Academy of Science). It took almost a quarter of a century before the same privilege was granted to a woman in the natural sciences (1971, Scientific Society of Finland).

Professorial appointments are made either by invitation or through open competition, where selection is to a great extent

49

determined by the ranking of referees (usually two or more) appointed by the universities. Since appointments by invitation are very rare, and usually there is considerable competition for professorial posts, the expert referees play an important role in the filling of tenured professorial posts. While in general the number of potentially eligible women referees (i.e. professors or associate professors) is relatively small, as the comparison of four different fields shows, when it comes to performing such an important task women remain invisible even in the female-dominated fields.

The marginalisation of women is even more evident in relevant societal decision-making bodies outside the intrascientific institutions. In the execution of its welfare function the state administrative machinery relies increasingly on the expertise of different interest groups. The institution of committees, subcommittees, boards, councils, etc. appointed by the state plays an important role in the machinery of the administration. These bodies have become a major forum for interest group influence. For example, in 1987, there were 468 committees and 676 subcommittees. Although the share of women in committees has steadily increased from 7 per cent in the mid-1970s to 15 per cent in 1987, their representation is still very low. Typically, women are rarely chosen to chair a committee but are often preferred as secretaries.

Thirty-two new committees were established in the year after the Law of Equality, stipulating the appointment of at least two women to a state committee, had come into effect. Of these, women members comprise 18 per cent and of chairpersons 6 per cent. Even now, 22 per cent of the newly established committees have not a single woman member. In the same year, 39 per cent of the subcommittees had no woman, the total proportion of women in subcommittees being 20 per cent, of chairpersons 10 per cent and secretaries 40 per cent. In the same vein, women comprised only 11 per cent of the external experts invited by the committees. (Silfverberg, 1988)

The extent of women's participation in national organisational structures concerned with the planning, direction and financing of scientific and technological activities is illustrated in Table 2.5.[10] The role accorded to women in these policy-making

10. The main S & T organisations in the public sector are as follows: The

bodies is an indication of their potential influence on the direction of the country's scientific activities. At the same time, the status accorded to membership in these bodies is a sign of recognition and, as such, constitutes an important source of capital of political and economic power for the participants. (Bourdieu, 1988)

In general the more direct the social and economic consequences of the activities of a given agency and the greater the influence of the private sector, the greater the invisibility of women.[11] This also seems to be the case for organisations

---

Science Policy Council is responsible for the overall planning, direction and co-ordination of research. Members are the Prime Minister (Chairman), five other ministers, the chairmen of the Central Board of Research Councils and the Council of Higher Education, and eight other members familiar with R & D, who may be appointed for a maximum of two terms of three years. The Academy of Finland, composed of a Central Board of Research Councils and seven subcommittees, is the main research funding agency in the public sector. The councils award research posts and grants. The members of the Central Board are the chairmen of the research councils and three other members appointed by the Council of State, and the chairman appointed by the President of the Republic. The research councils each have a chairman and from nine to fourteen other members, appointed by the Council of State for a maximum of two terms of three years. The Technology Development Centre plays a central role in the preparation and planning of technology policy and its implementation through technological R & D projects. The Centre's activities are supervised by a Board of Directors, appointed by the Council of State for a three-year term of office. The Council for Higher Education serves as an advisory committee to the Ministry of Education. The Council deals with overall development issues concerning institutions of higher education from the point of view of education as well as research. Appointments to some of these bodies are made upon consultations with the relevant organisations active in research. At least in the case of nominees for members to the Research Councils of the Academy of Finland and the Council for Higher Education, the nominating organisations have been especially requested by the Ministry of Education to take into account the equitable representation of women. This request so far has had no significant effect.

11. The absence of women from influential positions is especially notable in the case of the major economic organisations. The combined total membership of the three major trade union organisations, the Central Organisation of Finnish Trade Unions, SAK; Confederation of Salaried Employees in Finland, TVK; Confederation of Technical Employee Organisations, STTK, comprised in 1981 two thirds of the labour force. The proportion of women union members ranged from 14% (STTK) to 43% (SAK) and 81% (TVK). The corresponding representation of women in the executive committee of the respective union organisations was 4%, 8% and 31%. Women are even less represented in employers' organisations. Only one of the 28 member organisations of the Finnish Employers' Confederation has any women in its executive committee. Nor did any of the 18 co-operative concerns investigated by the Council for Equality have a single woman on their board (Position of Women, 1984).

Veronica Stolte-Heiskanen

Table 2.5: Participation of women in the political and economic power structures of science and technology

Organisation	Number of Women Total	In sub-sections	Total number of members
Science Policy Council	2		16
Academy of Finland			
Central Board of Research Councils	2		11
Research Councils	23		105
Humanities		5	15
Natural sciences		3	15
Medical sciences		4	15
Agriculture and forestry		2	15
Technical sciences		2	15
Social sciences		3	15
Environmental sciences		4	15
Council for Higher Education	18		102
Section chairpersons	3		11
Central council members	3		12
Section for:			
Humanities and theology		2	11
Education and psychology		4	10
Teacher training		2	13
Business economics		0	8
Social sciences		2	11
Natural sciences		2	11
Technical sciences		0	6
Medical sciences		1	11
Arts		1	5
Higher education research		4	9
Computer sciences		0	7
The Technology Development Centre, TEKES			
Board of directors	1		9
(alternate members)	0		9

Table 2.5 *continued*

Organisation	Number of Women Total	In sub-sections	Total number of members
Advisory board	0		17
(alternate members)	1		17
Scientific Committee of National Defence	0		45
Governing or advisory boards of state research institutes			
Agricultural Research Centre	0		6
Economic Planning Centre	0		4
Forest Research Institute	0		8
Game and Fisheries Research Institute	0		6
Geodetical Institute	1		8
Geological Research Institute	0		9
Institute of Marine Research	0		5
Institute of Occupational Health	2		13
Law Policy Research Institute	1		9
National Institute for Dairy Research	0		7
Technical Research Centre of Finland	0		11
National Public Health Institute	0		5
National Research Institute of Farming Machinery	0		8
Peace and Conflict Research Institute	2		11
Research Centre for Domestic Languages	1		7

continued on page 54

Table 2.5 *continued*

Organisation	Number of Women Total	In sub-sections	Total number of members
Research Institute for Agricultural Economics	0		7
Water and Environment Research Institute	2		5
Finnish National Fund for R & D	0		9
R & D supporting private foundations (N = 47)		•	
Chairperson and other leading functionaries	1		67
Secretaries	7		84

Source: Suomen Valtiokalenteri, 1989; annual reports and/or communications with relevant agencies.
Data are for 1987 or 1988.

concerned with science and technology. For example, women are conspicuously absent from the decision-making bodies of the private research financing foundations. While the share of private foundations in the overall R & D investments of the country is very small, from the point of view of the individual scientist a foundation grant may be decisive for the realisation of a research project.

The Ministry of Education and the Ministry of Trade and Industry are the two main ministries in science administration. As recently as 1987, of the 59 committees of the Ministry of Trade and Industry, 8 per cent of the total of 717 members and 7 per cent of the chairpersons were women. At the same time almost one half of the committees did not have a single female member. Correspondingly, of the 108 committees of the Ministry of Education, 26 per cent of the total of 1,203 members and 14 per cent of the chairpersons were women. Only 11 per cent of the committees did not have a single woman member. Some recent examples of all-male committees important from the point of view of science and technology are: the Ministry of

Trade and Industry's subcommittee on 'Science and Technology of European Countries', the Ministry of Education's subcommittee, 'Development of Research in Economics' or the Ministry of Foreign Affairs' 'Finnish-Soviet Scientific-Technological Cooperation' committee. (Silfverberg, 1988)

Conclusions

Although the participation of women in science has increased during the last decades, women on the whole occupy a peripheral position in Finnish science. There are not only few women in higher level academic and research positions, but they are even more notably absent from positions of influence related to the organisation and orientation of scientific activities, knowledge production and its utilisation. On the structural levels, women scientists' cumulative disadvantage is reflected in the reproduction of the typically pyramidal shape of the position of women in the hierarchical structure within institutions across different levels of the power structures of science and society. The closer we get to the politico-economic power structures, the steeper the pyramid becomes.

Cultural, social and motivational factors undoubtedly play a role in keeping women either on the margins or totally away from science. Gender differences in scientific orientations are already evident on the secondary school level even in such a gender-equality oriented society as Finland. As recently as 1986, boys chose the advanced mathematics alternative twice as often as girls, and one out of every five girls, but every other boy chose advanced physics in their curriculum. (Komiteanmietintö, 1988) Consequently, by the time girls and boys are ready to embark on their university studies, orientation towards science and technology becomes one of the most distinguishing gender-linked factors determining educational choice. As Häyrynen notes (1984), students' self-perceptions about their professional abilities, combined with the climate of the faculty, lead to tendencies to make choices which conform to traditional academic and professional images. For women this means that they are channelled into typically female fields as well as other solutions conforming to feminist motivational stereotypes.

In the process of many years of gender socialisation women

are still discouraged from internalising such 'unfeminine' qualities as independence, initiative and assertiveness – the conventional personality characteristics of successful scientists. The absence of role models with whom women can identify and the lack of encouragement from teachers and colleagues further reinforce stereotyped self-images and feelings of inadequacy. A significant minority of women scientists consider lack of encouragement as the reason for women's marginal status in science. (Stolte-Heiskanen, 1982; Bergman, 1985) Thus, the encouragement of women to choose scientific careers must begin during the formative stages of the individual's personality, on the one hand, and with the demasculinisation of cultural images of a typical scientist, on the other. This would require more active measures to enlist the support of the family, the educational and scientific system and the mass media.

As the foregoing analysis has shown, simple macro-level participation rates in higher education and in the scientific labour force alone are poor indicators of the actual status of women in science. They only conceal the variety of social, cultural and political mechanisms that produce the significant qualitative differences in the typical career opportunities of men and women scientists. For example, Finnish women have one of the highest rates of participation in higher education. However, for women higher education is a necessary but not a sufficient prerequisite for upper white-collar occupations, such as science. Contrary to common expectations, the substantial gains in women's access to higher education have not led to better career opportunities for women.

According to a 1986 survey, while among those born between 1930 and 1939 gender does not differentiate university graduates' chances of getting upper white-collar positions, in the 1950-59 age cohort a considerably lower proportion of women graduates than men were in the upper class occupations. (Pöntinen, 1988) Thus the realisation of educational equality was accompanied by lower occupational chances for women.

Education is obviously not as good a resource in the labour market for women as it is for men. One can hypothesise that this is due to the changing role of higher education. With the structural transformation of Finnish society the knowledge class is gaining increasing significance, and higher education is becoming more an instrumental than a purely cultural resource.

Alongside the economic elite, the dominantly male scientific and educational elites have gained major importance during the last decades. (Haavio-Mannila, 1981)

With the growth of neocorporativism typical of the Nordic welfare states, policy decisions are strongly affected by the influential corporative organs of emergent interest groups. The strategic politico-economic power structure is increasingly articulated through a network of committees, commissions, councils, etc., composed of administrators and representatives of strategic organisations and experts drawn increasingly from the knowledge class. In this system women are clearly either totally excluded or at best under-represented.

In the social arena of science too, it is evident that women have a very limited access to structures of scientific power and prestige – and even to a greater extent – to political and economic power. The invisibility and marginalisation of women in these bodies has the double consequence of, on the one hand, excluding women from having influence on the direction of the cognitive and social development of scientific activities and of, on the other hand, further promoting the reproduction of male monopoly over scientific credit.

According to the recent OECD evaluation of Finnish science policy, 'TEKES (The Technology Development Centre) represents today in Finland the pivot of technology policy and holds with the Academy of Finland on the science side many of the keys to the national future' (OECD, 1987:90). This conclusion eloquently illustrates the broad social significance of these and other science policy-making bodies. If visions of the 'national future' are to include the equitable participation of women, the much publicised Law of Equality will have to be consciously realised not only in principle but also in practice.

Appendix: The Academic Teaching Posts and Research Fellowships of the Academy of Finland (S.A.)

Position	Degree requirement	Terms of appointment	Obligations
Senior level			
Research professor (SA)	Ph.D.	3–5 years of renewable terms, permanent	Research occasionally teaching
Full professor	Ph.D.	Tenure	Teaching and research
Associate professor	Ph.D.	Tenure	Teaching and research
Senior researcher (SA)	Ph.D.	3 years, 2 consecutive 3 year renewal possible	Research
Middle level			
Lecturer	Licentiate	Tenure	Teaching
Senior assistant	Licentiate	5 years, renewable	Teaching and research
Junior researcher (SA)	Licentiate	3 years, 2 consecutive 3 year renewable possible	Research
Lower level			
Faculty assistant	MA	5 years, renewable	Teaching and research
Research assistant (SA)	MA	3 years, 2 consecutive 3 year renewable possible	Research

References

ANTIKAINEN, A. AND A. JOLKKONEN, *Sivistyneistöä vai teknokraatteja? Selvitys tohtoriksi ja professoriksi tulosta Suomen korkeakoululaitoksessa historiassa, kasvatustieteessä, liiketaloustieteessä ja sosiologiassa vuosina 1950–85* (Intellectuals or technocrats? Becoming a professor in history, education, business economics and sociology), Tampereen yliopisto, Kasvatustieteen laitos, Tutkimusraportti A 43, Tampere, 1988.

BERGMAN, S., *Kvinnan och vetenskapssamfundet – teoretiska bidrag till förståelsen av hennes utbildningsbeteende och en explorativ studie bland forskarstuderande vid Åbo Akademi* (Woman and the scientific community. A study of post-graduates at Åbo Akademi), Naistutkimusmonisteita 1985: 1, Tasaarvoasiain neuvottelukunta, 1985.

BOURDIEU, P., *Homo academicus*. Cambridge: Polity Press, 1988.

BRUUN, K., K. ESKOLA, AND K. SUOLINNA, Väitöksestä juhlakirjaan: naiset tieteen miesseuroissa (From dissertations to Festschrift: Women in the men's clubs of science). *Sosiologia*, vol. 19, no. 2, 1982: 89–101.

COLE, J.R., *Fair science: Women in the scientific community*. New York: The Free Press, 1979.

DISCO, C., 'Intellectuals in advanced capitalism: capital, closures, and the new class thesis', in R. Eyerman, L.G. Svensson & T. Söderquist (eds), *Intellectuals, universities and the state in Western modern societies*. London: University of California Press, 1987, pp. 50–77.

FERRY, G. AND J. MOORE., 'True confessions of women in science'. *New Scientist*, 95, 1981, pp. 27–30.

FOUCAULT, M., *Language, counter-memory, and practice. Selected essays and interviews*, ed. D. Bouchard. Ithaca, N.Y.: Cornell University Press, 1977.

GOULDNER, A., *The future of the intellectuals and the rise of the new class*. New York: Macmillan, 1979.

HAAVIO-MANNILA, E., 'Women in the economic, political and cultural elites in Finland', in C.F. Epstein and R.L. Coser (eds), *Access to power: Cross-national studies of women and elites*. London: George Allen & Unwin, 1981, pp. 53–75.

——, 'Nainen "miesten töissä": naisinsinöörinä Suomessa' (Women in 'men's jobs': female engineers in Finland), in H. Varsa (ed.): *Naiset, tekniikka ja luonnontieteet* (Women, technology and natural sciences). Tasa-arvoasiain neuvottelukunnan monisteita 8/1986, Helsinki: 1987, pp. 14–22.

HALILA, A., *Suomalainen tiedeakatemia 1908–1983* (The Finnish Academy of Science, 1908–1983). Porvoo: WSOY, 1987.

HASSI, S., *Naiset ja tekniikka* (Women and technology). Tasa-arvoasiain neuvottelukunnan monisteita 6/1986.

59

HÄYRYNEN, L., 'Koulutetun naisen elämänkulku' (Life patterns of educated women). *Naistutkimus*, 1, 1988, pp. 3–18.

HÄYRYNEN, Y-P., 'Suomalainen korkeakouluopiskelija' (Finnish university students), in T. Ikonen, J. Jussila, and K.E. Nurmi (eds), *Korkeakouluopetuksen teoriaa ja käytäntöä* (Theory and practice of higher education). Helsinki: Opetusministeriö, 1984, pp. 27–75.

JULKUNEN, R., 'Jyväskylän yliopiston naisopettaja' (Women academics at Jyväskylä University), in '*Komiteanmietintö 1982:33 Liite C: Naisten tutkijanuran ongelmat ja esteet. Opetusministeriön asettaman työryhmän mietintö*' (Committee Report of the Ministry of Education on the problems of women's scientific careers 1982:33 Annex C). Helsinki: 1982.

KIRKHAM, K. AND P. THOMPSON, 'Managing the diverse work force: women in engineering', *Research Management*, 27, 1984, pp. 9–16.

Komiteanmietintö 1982:33, Naisten tutkijanuran ongelmat ja esteet. Opetusministeriön asettaman työryhmän mietintö. (Committee Report of the Ministry of Education on the problems of women's scientific careers 1982:33). Helsinki, 1982.

Komiteanmietintö 1988:17. Tasa-arvokokeilutoimikunnan mietintö (Report of the committee on experiments in equality 1988:17). Helsinki: Opetusministeriö, 1988.

Korkeakoulut 1986 (Higher education 1986). SVT. *Koulutus ja tutkimus 1988:5*. Helsinki: Tilastokeskus, 1988.

KOSKIALA, S., *Finnish women in engineering. Unesco consultative meeting on women in engineering and technological education and training*, Paris, 2–4 December 1986, Final Report. Paris: Unesco, 1986, pp. 49–53.

LIPMAN-BLUMEN, J., 'Toward a homosocial theory of sex roles: an explanation of the sex segregation of social institutions', *Signs*, vol. 1, no 3, 1976, pp. 15–31.

LUUKKONEN-GRONOW, T., 'University career opportunities for women in Finland in the 1980's', *Acta Sociologica*, vol. 30, no 2, 1987, pp. 193–206.

——, 'Women's research career – obstacles and incentives', in E. Kaukonen and V. Stolte-Heiskanen (eds): *Science Studies and Science Policy*, Publications of the Academy of Finland 3/1984. Helsinki, 1984, pp. 161–82.

——, AND V. STOLTE-HEISKANEN, 'Myths and realities of role incompatibility of women scientists', *Acta Sociologica*, vol. 26, no 3/4, 1983, pp. 267–80.

MORGAN, D., 'Men, masculinity and the process of sociological inquiry', in H. Roberts (ed.), *Doing feminist research*. London: Routledge & Kegan Paul, 1981, pp. 83–113.

OECD, *Reviews of national science and technology policy: Finland*. Paris: OECD, 1987.

Population Census, Volume 1, Economic activity of population. Official Statistics of Finland. Helsinki: Central Statistical Office of Finland, 1988.
Position of women. Statistical surveys No. 72. Helsinki: Central Statistical Office of Finland, 1984.
PÖNTINEN, S., *Development of educational opportunities in Finland*. Paper presented at the Second Finnish–Hungarian Sociological Seminar, Budapest, September 5–10, 1988.
RESKIN, B.F., 'Sex differentiation and the social organization of science', in J. Gaston (ed.), *The sociology of science*. San Francisco: Jossey-Bass, 1978.
ROSSITER, M.W., 'Sexual segregation in the sciences. Some data and a model', *Signs*, vol. 4, 1978, pp. 146–51.
RÄTY, T. AND LUUKKONEN-GRONOW, T., *Korkeakoulujen opettajakunnan vaihtuvuudesta ja ikärakenteesta vuosina 1967–79* (The age structure and turnover of university teachers during 1967–79), Suomen Akatemian julkaisuja 12/1981. Helsinki, 1981.
SAARI, S. AND H. MAJANDER, *Nuorten akateemisten työntekijöiden työolot ja mielenterveys*. (Work conditions and mental health among young academic employees), University of Helsinki, Department of Psychology, Research Reports No. 1. Helsinki, 1985.
Seurantatyöryhmän mietintö, *Naisten tutkijanuran ongelmat ja esteet*. (Report of the Follow-up Subcommittee: problems and obstacles of women research careers), Opetusministeriön työryhmien muistioita, 1986:34. Helsinki, 1986.
SILFVERBERG, A., *Naiset komiteoissa* (Women in committees), Sosiaali-ja terveysministeriö, Tasa-arvojulkaisuja, Sarja C, Työraportteja 2/1987. Helsinki, 1988.
SILIUS, H. (ed.), *Kvinnor i mansdominerade yrken* (Women in male-dominant professions), Publikationer från Institutet för Kvinnoforkning vid Åbo Akademi, nr. 5. Åbo, 1989.
STOLTE-HEISKANEN, V., 'Tieteen universalismi ja sukupuoliroolit: miehet ja naiset tutkimuksen arkielämässä' (Universalism of science and sex roles: men and women in everyday life of research), *Sosiologia*, vol. 18, no 1, 1981, pp. 13–23.
——, 'Naisen tutkijanuran ongelmat: naisen vai tutkijan ongelmat?' (Problems in women's scientific careers: women's problems or scientists' problems?) In *Komiteanmietintö 1982:33 Liite B: Naisten tutkijanuran ongelmat ja esteet*, Opetusministeriön asettaman työryhmän mietintö (Committee Report of the Ministry of Education on the problems of women's scientific careers 1982:33 Annex B). Helsinki, 1982.
——, 'The role and status of women scientific research workers in research groups', in J.H. Pleck and H.Z. Lopata (eds), *Research in the*

interweave of social roles: families and jobs, 3, JAI Press, 1983, pp. 59–87.
Suomen valtiokalenteri 1989 (State register of Finland 1989). Porvoo, WSOY, 1989.
WESTOBY, A., 'Mental work, education, and the division of labor', in R. Eyerman, L.G. Svensson and T. Söderquist (eds), *Intellectuals, universities and the state in Western modern societies*. London: University of California Press, 1987, pp. 127–53.
VETTER, B.M., E.L. BABCO AND J.E. MCINTIRE (eds), *Professional women and minorities. A manpower data resources service*. Washington: Scientific Manpower Commission, 1978.
ZUCKERMAN, H. AND J. COLE, 'Women in American science', *Minerva*, vol. 15, no 3, 1975, pp. 82–102.

3

Women in Science Careers in the German Democratic Republic[1]

Heidrun Radtke

Women's Entry into the Realm of Science and Technology

Women in the German Democratic Republic (GDR) realised their professional potential in the realms of science and technology relatively late. The process began in the latter half of the twentieth century – at a time when women working in industry and manufacturing, in health and education services and in administration had already become an important part of the labour force.

In 1908, almost fifty years later than in other countries, official permission was finally granted to women who wanted to study at university. The right to teach at a university and to obtain a doctorate was granted in 1920 within the context of the political rights (i.e. those of suffrage) gained as a result of the November Revolution in 1918. Whilst women in manufacturing and administrative positions had already become the norm by the middle of the twentieth century, they were only just beginning to develop their intellectual and creative potential within social sciences and achieve academic acclaim for their work. In the realm of natural science and technology this process began even later. For a variety of reasons, women from a secure, middle-class background had no reason to enter into employment. However, women from this social class were the only women able, from both a financial and educational point of view, to take advantage of an intellectually more demanding occupation

---

1. The author carried out her research during 1988 and 1989, in the GDR, before the country was re-united.

requiring systematic exposure to academic problems, which, in turn, could lead to a certain degree of professionalism. They chose an academic vocation because they themselves were interested in intellectual problems and had a personal need to realise their true potential through their work. This was seldom in accordance with the interests of the family and the demands of motherhood, due to a lack of support within society to ease the burden of the task of bringing up children and attending to domestic chores.

The women academics of the time therefore had to battle against enormous social obstacles. Indeed, from the very start of their entry into a profession women have had to contend with prejudices arising from doubts surrounding their academic capabilities. In daily life and in science, theories had arisen which questioned the ability of women to work on an intellectual level, to think logically or to operate effectively in an academic environment because of the alleged incompatibility of motherhood and feminity with intellectual work. Philosophers, biologists, physicians, and especially anthropologists, advanced various theories allegedly demonstrating the inferior intellectual ability of women.[2] These theories are even today a source from which long-established prejudices can be drawn when talking of the intellectual capacity of women, and thus this still constitutes a determining factor in influencing society's opinion of women working in science.

As part of the general strategy of equal rights for men and women, the GDR nurtures a milieu which allows women to combine a vocation with the demands of motherhood, thus making more careers available to them. This applies especially to the realm of science and technology and positions of responsibility within them. At the centre of such a milieu there is suitable legislation which allows women to combine a career with being a mother with a comprehensive support package. In the GDR women are encouraged to have children as well as to obtain qualifications at levels of higher and technical education. For this reason special arrangements can be made between women and the heads of academic institutions, trade union representatives and youth organisations.

2. The classical reasoning used to explain the different intellectual capacities of men and women by the neurologist Paul Möbius (1853–1907) – based on the relative size of the brain – has been proven wrong.

In science it is principally the female students and research assistants with children who are encouraged to achieve their academic goal through individual timetables, financial help and preferential treatment when it comes to places in day care centres for their children. Upon the basis of different sociopolitical measures young women can already realise their desire for children during the time of their studies which need not be interrupted in the case of pregnancy. In the case of a research assistantship, employers are obliged to ensure that the necessary steps are taken to enable the woman to fulfil her scientific obligations. Practical measures include the placement of the children in communal or work-based nursery schools or kindergartens, providing medical assistance, as well as guaranteeing the woman a suitable position within the firm on the successful completion of the research course.

Against this background the number of women entering higher education in the GDR since its foundation in 1949 steadily increased. The share of women among university students increased from about one quarter in 1950 to over one half in 1987. In 1987, 53 per cent of the graduates were women. Consequently there is now a large qualified female potential S & T manpower and thus no reason to exclude women from positions of responsibility in science and technology.

Women are generally well represented in the social sciences, educational sciences, economics and literary studies or languages. However, increasingly more women are making their presence felt in the traditionally male dominated environment of the natural sciences.

In the natural sciences women are most strongly represented in chemistry, biology, psychology, and pharmacy. The number of women studying engineering and technology has also steadily increased, although not as quickly. Today the proportion of women is 30 per cent, although their share is much larger at technical and engineering colleges. Within the engineering disciplines women are mostly to be found in subjects such as material science, manufacturing engineering, town planning, architecture and data processing (more than half the students in these subjects are women).

Table 3.1: The proportion of women in higher education according to fields of study, 1987

Field	% women students
Natural Sciences	50.8
Engineering and Technical Sciences	27.5
Medical Sciences	56.6
Agricultural Sciences	47.2
Economics	67.7
Philosophy/History/Political Science Jurisprudence	35.6
Sports	40.8
Literary Studies/Languages	60.5
Art	44.2
Educational Sciences	73.0

Source: *Colleges of Higher and Technical Education in the GDR – a Statistical Account.* Berlin: Ministry of Higher and Technical Education, 1988.

Present Trends in Women's Participation in Positions of Responsibility in Science[3]

Today a relatively larger percentage of women graduates opt for an academic career. Currently the more than 20,000 women represent 43 per cent of the academic staff of universities and colleges (1988). Since 1975, there has been a continual increase in the proportion of women in the academic staff in all institutions of higher education in the fields of culture, education, agriculture, forestry and food processing.

3. Under 'positions of responsibility in careers of science' are meant women in leading academic positions, e.g. female professors, lecturers and heads of scientific groups or collectives. Women in 'leading positions' are women who lead academic teams, for instance heads of departments. Women in 'top positions' are women who lead big academic institutions; these are rectors of universities or directors of institutes or sections in academies, universities and other higher educational institutions.

A leading position or other positions of responsibility in science can be obtained through 'Promotion' (procedure followed leading to the acquisition of a higher academic qualification) in two steps. The first step is the *Promotion* A = for a Doctor of a branch of science ('Dr'); *Promotion* B = for a Doctor of Science ('Dr sc.'), earlier known as 'Habilitation' ('Dr habil'). This is the basis for higher academic titles in science; lecturers and professors are the highest degrees.

In the natural, engineering, and technical sciences at the moment, the average growth in the number of women entering the academic body of technical and engineering colleges is greater than that of the number of female students. The same is true within the association of engineers, the Chamber of Technology, which brings together not only academic engineers but also those from many other areas (including the manufacturing industry). The president of this organisation is a distinguished woman scientist.

The quality as well as the quantity of potential female academic recruits has improved. Between 1976 and 1986 the number of women having obtained a Doctorate (Promotion A) rose from 21 to 32 per cent and of those obtaining an advanced doctor's degree (Promotion B) from 8 per cent to 15 per cent. In 1986 twice as many women took their Doctorate as in 1976, including in engineering and natural sciences. Compared with the number of women obtaining a Doctorate in the educational sciences, literary studies, languages and economics, where their share is more than average, the proportion of women in natural sciences and engineering is still relatively low. (Radtke, 1988)

Despite the large number of qualified women, it seems that the number of women who are qualified to hold leading positions is not reflected in the number of women who actually do so. This is especially the case in disciplines where the share of women among potentially qualified scientists is more than 40 per cent. No academic institution has more than 15 per cent of its decision-making positions filled by women. At universities and colleges 12 per cent of the lecturers, 5 per cent of the professors and barely 3 per cent of the chancellors, vice-chancellors and heads of department are women. The proportion of women occupying positions of middle-level leadership at such institutions is 10 per cent.

For some years the percentage of women occupying positions in science (albeit still small in absolute numbers) has witnessed a more rapid rate of increase than that of the whole female scientific potential. The access of women to leading positions, however, has not kept pace everywhere with the development of female scientific potential as a whole. There are also differences between institutions. At several research institutions the percentage of women professors is lower than the national average. Above all there is everywhere a lack of women in top-level positions. Here the share of women amounts to 2.6 per cent, which is much less than the number of women professors or women scientists with advanced doctorates, i.e. those qualified for leading positions. Thus a lot of opportunities still continue to be unused concerning an effective utilisation of qualified female scientific manpower. As a result, women's presence in positions of responsibility presents the picture of a pyramid which, when applied to men, is turned upside down, i.e. has a relatively narrow base.

Equal opportunities for men and women concerning positions of responsibility would mean that the distribution of women in different positions would eventually have to resemble a cube rather than a pyramid.

The share of women in positions of responsibility also varies among different scientific disciplines. There is scarcely a field with a well-balanced proportion among women on different hierarchical levels.

In order that more women might find their way to positions of responsibility in careers of sciences and technology (and in particular to top positions) a selective encouragement of new academic talent is necessary, especially with regard to the acquisition of an advanced doctorate. An advanced doctorate is the necessary precondition for positions of leadership and supervision in science; it is not an end in itself but prepares the academic for more responsible tasks and leadership of larger scientific groups. It is an important step on the way to the highest academic title and to a position at the top of science. At the moment, however, the number of women taking their advanced Doctorates is not consistent enough to ensure that enough qualified women academics will be available in the future to fill higher positions. It is therefore necessary to choose, from the relatively large number of female students, those especially suited and motivated, and encourage them on to a course of more demanding scientific work. This would make it possible for more women to achieve the highest academic qualification possible.

The causes of the partly unsatisfactory state of affairs regarding the placement of suitable women candidates in leading positions are various and manifold. They are both objective and subjective. A certain level of academic success and ability must be achieved at work in order that a position of leadership may be justified – something which is often difficult for a mother in full-time employment to achieve. Nevertheless, sociological studies have shown that in the GDR the two can generally be combined and that the main causes for not achieving are primarily subjective ones. (Radtke, 1988)

Prejudices Concerning the Ability of Women

One of the biggest obstacles preventing a larger number of women from holding a decision-making position within science – especially top positions – is the still prevalent doubt about the intellectual ability of women. It is taken for granted that women are less suited for conceptual and scientific work and that a woman with children cannot function in an executive position and carry out her scientific tasks with above average success. As a consequence of these assumptions women are automatically expected to achieve less than they are able to. This considerably reduces any real chance they might have of holding a position of responsibility in the future. Checks made by the state through the 'Workers' and Farmers' Supervisory Committee' to ensure that the laws are observed have repeatedly shown that a person's sex is a more important issue than the professional suitability of the candidate for a job. Even some of the best women students had experienced difficulties in getting a job because of the number of children they had.

Inadequate Academic Utilisation of Women

Successful scientific work requires high expectations of oneself and of one's work, self-confidence, self-assertion and a particular desire to do well. The characteristic of successful women in science is not that they are working under more comfortable conditions than other women, but that they can work independently and profitably, and are able to develop concrete ideas regarding career and family. They are thus able to organise their time at work and at home around the realisation of these aims.

In order to remove the prejudices about the capabilities of women academics and in order to encourage their access to positions of responsibility, women must be seen to maintain a high level of achievement over a long period of time. A demanding research project is, for this reason, the most important way to encourage women in this respect. This means a project which objectively harbours the possibility of achieving a breakthrough. However, it is far more usual for men to have access to such projects than for women. This is especially true of projects in the field of science and technology, which could lead to patentable solutions, by virtue of which scientists can feel their own ability being tested and which represent internationally high standards.

'Respect' for the family obligations of very able young mothers by keeping them away from demanding research projects has worked to the disadvantage of such women. In order that women academics might continue their career advancement it is necessary for women to receive demanding tasks at the beginning of their careers. This will avoid the fallacy that the demands placed on the woman in her home environment become less as the children become older and more intensive scientific work only then becomes possible. Such assumptions contradict the nature of scientific work. It is a profession in which the climb to the top is a long one and early exposure to a certain amount of responsibility is a prerequisite for the successful fulfilment of the future role of a director.

The Compatibility of Motherhood and a Career in Science

A high degree of professional commitment, thorough preparation, academic excellence and the willingness to devote a lot of free time to scientific work are necessary to achieve positions of responsibility in science. For a woman with a family an enormous amount of effort is needed to realise the demands of her professional goals, all the more so as each of the demands is a challenge in its own right. The combination of a career and of motherhood, the demands of a leading position and the specific nature of the scientific task itself, all demand complete dedication if a desired level of achievement is to be attained.

In order to make it possible for women to lead successful careers under these conditions without having to sacrifice their self-realisation within the family, a social milieu has systematically been developed on a long-term basis (and is still being developed) in order that this difficult aim might be achieved.

Social institutions responsible for the care and education of children as well as socio-political provisions allowing for the combination of a career and motherhood are at the heart of such a milieu. Women in the GDR are materially and socially secure. They are not dependent on men and are financially self-supporting. Against this background and given that there are now a large number of highly qualified women – 54 per cent of all university and technical college graduates are women – it is now possible, indeed necessary, for women to strive for decision-making positions within their chosen career.

In order that women might play a greater role in positions of responsibility and leading positions, the necessary social conditions must continue to be improved so that women do not have to choose between a career and motherhood. For this reason such a policy has become an integral part of the economic and social policy of the government. Social security and the combining of a career and public duties with motherhood are part of the national policy and are actively supported by the trade unions. The policy allows women in managerial positions and women who have a relatively long period of training ahead of them – as is typical of women academics – to continue without having to sacrifice family life. Indeed, most women in leading positions do have children, or rather a family. Most women university professors and lecturers have two children, which corresponds to the national average. From studies of female academics, students and engineers it can be concluded that motherhood is in no way incompatible with a high level of achievement or a leading position. (Radtke, 1988; Waltenberg, 1989) Moreover, women with children are capable of achieving as much as women without children and even show a level of achievement higher than the average.

Women who themselves hold an executive post recognise, on the basis of their own experience, that the family is not to be regarded as a hindrance. Opposite views, in which the ability of a woman to assume a leading position is seen as an alternative to motherhood, are held predominantly by men or by those women who either hold no such position or have no family.

Women academics in positions of responsibility have achieved their professional standing by virtue of particular attributes of their character and an agreement between themselves and their partners regarding career and aims in life, rather than because of particularly suitable circumstances or because they have chosen to do without a family. Decisive factors dictating the entry of women into executive positions or for a successful career include the motivation to do scientific work, the ability to come to terms with difficulties and obstacles in one's personal development, and the ability and willingness to assume a position of responsibility. The desire to succeed in life in general underlies these characteristics along with a determination to achieve personal and professional goals. Women in possession of such qualities develop strategies – often down to a fine art – which enable them to combine their personal and professional activities, or rather the demands of their position with those of their family, so that the one does not necessarily exclude the other, or any contradictions arising out of the situation can be overcome.

A general increase in the demands women place on work, on a relationship and the family, on social activities and relationships, calls for new conditions which allow women in science to better combine decision-making with the realisation of their role in society as mothers. The strategy should therefore continue to be one of society providing increasingly more consumer services – those for which in the main the family today still has to provide – and the continuing development of objective and subjective conditions to allow fathers to better accomplish their responsibilities to the family.

References

BERTRAM, B., 'Junge Frauen in leitende Tätigkeiten – Voraussetzungen nutzen' (Young women in decision-making positions – using preconditions), in *Informationen des Wissenschaftlichen Rates 'Die Frau in der sozialistischen Gesellschaft'*, the Academic Council 'The woman in socialist society' of the Academy of Sciences of the GDR, vol. 5, 1987, S. 3–19.

BURRICHTER, G. (ed.), *Theorie und Praxis der Wissenschaftsforschung* (Theory and practice of scientific research). Praxis und Konzepte, Institut für Gesellschaft und Wissenschaft. Erlangen, 1987.

'Die Frau – der Frieden und der Sozialismus. Kommunique des Politbüros des Zentralkomitees der SED vom 23. Dezember 1961' (The woman – Peace and Socialism. Communiqué of the Politbureau of the Central Committee of the Socialist Unity Party of Germany, 23.December 1961), in *Die Frau – der Frieden und der Sozialismus*, Central Committee of the SED, 1962, S. 3–35.

FEYL, R., *Der lautlose Aufbruch. Frauen in der Wissenschaft* (The Silent Awakening. Women in Science). Berlin: Verlag Neues Leben, 1981.

GDR REPLY TO THE UN SECRETARY GENERAL. 'Zahlen unfd Fakten zur sozialen Sicherheit in der Deutschen Demokratischen Republik' (Figures and Facts on social security in the German Democratic Republic). Berlin: News Agency Panorama DDR, 1988.

Gesellschaftliche Entwicklung der Frau – Vereinbarkeit von Berufstätigkeit und Mutterschaft – Demographische Prozesse – Frauenforschung – Information und Dokumentation (Social Development of women – Compatibility of career and motherhood – Demographic processes – Research about women – Information and documentation). 3rd International Demography Seminar, Institut für Soziologie und Sozialpolitik der Akademie der Wissenschaften der DDR. Forschungsgegenstand: Die Frau: S. II–IV. Berlin, 1988.

'Gesetz über den Mutter- und Kinderschutz und die Rechte der Frau' (Act concerning the protection of mother and child and the rights of women), in *Frauen schaffen für das neue Leben*. Berlin: Dietz Verlag, 1955.

Ministry of Higher and Technical Education, *Colleges of Higher and Technical Education in the GDR, A Statistical Account*. Berlin, 1988.

RADTKE, H., 'Frauen in Leitungsfunktionen der Wissenschaft' (Women in leading positions in science), *Einheit*, vol. 10, 1988, pp. 930–36.

STOLTE-HEISKANEN, V., 'Rolle und Status von Wissenschaftlerinnen in Forschungsgruppen' (The role and status of women academics in research groups), in *Informationen des Wissenschaftlichen Rates 'Die Frau in der sozialistischen Gesellschaft'*, the Academic Council 'The woman in socialist society' of the Academy of Sciences of the GDR, vol. 1, 1985, S. 21 f. Berlin.

WALTENBERG, CH., 'Nicht das Kind und nicht der Haushalt . . . Frauen in der Wissenschaft' (Not the child and not the housework . . . women in science), in *Sonntag*, 19, 1988, Berlin.

——, 'Forschungsgegenstand: Die Frau' (Research subject: The woman), in *Spectrum*, the Academy of Sciences of the GDR, vol. 11, 1989.

Zur gesellschaftlichen Stellung der Frau in der DDR (The social status of women in the GDR). Leipzig: Verlag für die Frau, 1978.

4

Double-Faced Marginalisation. Women in Science in Yugoslavia

Marina Blagojević

Introduction

The position of women in science is a result of the position of women in the given society, on the one hand and the position of science, on the other. This seemingly simple statement belies the many complex relations that exist between gender, social status, social promotion, openness of the social structure, education, social development in general and the development of science (in its cognitive, theoretical, methodological aspects and in its organisational aspects), as well as family life and its main characteristics.

Women in Yugoslavia are a marginal group. Although they enjoy all the legal rights (since the Socialist Revolution) they do not have the same social status as men. The more prestigious the area of social life, the more closed it is to women. Politics is the most closed of all.

Science itself is in an ambivalent position. Although verbally and ideologically treated as one of the 'wheels of development', the real position of science in Yugoslavia deteriorated, climaxing in the present deep socio-economic crisis. As it became less and less prestigious science increasingly opened up to women.

This paradox deeply questions the relationship between social development and the improvement of women's position in science. On the one hand, as can easily be proved by comparing different regions in Yugoslavia,[1] social development for women

---

1. In this paper 'region' is used to denote all the republics (Bosnia and Herzegovina, Montenegro, Croatia, Macedonia, Slovenia, Serbia) and two provinces (Kosovo and Vojvodina). Because of very great differences between

means better education, higher employment rates, and consequently better positions in professions, including science. On the other hand, factors that actually impede development, such as irrational growth of the educational system which produced many highly educated unemployed people, the low position of science, and intensive 'brain drain' (especially men's), produced better chances for women under worse conditions. The gain of women should, in this case, be evaluated only relatively.

Basic Theoretical Premises

Of the many relevant approaches to the problem at hand, this paper focuses above all on the sociological aspects. The first step is to try to distinguish and structure the relevant social factors. They operate at three levels: on the level of the societal system, on the level of science as a social subsystem and on the level of the individual, the woman scientist herself.

Operating on the first, societal level are the factors which condition the position of women as a social group. This means the characteristics of the class-strata, social structure, the characteristics marking the degree of economic development (mode of production) and the characteristics of the dominant social ideology (the dimension of egalitarianism/elitism).

On the 'societal level' women are a marginal group with fundamental ideological and economic functions. Economic, because women provide a labour force for lesser paid and less prestigious occupations (e.g. farming, semi-professions),[2] as well as a labour force for biological reproduction and the daily support of members of the household (child care and housework). On the ideological level, women as a marginal group make it possible to maintain the social structure's status quo. As they belong to different social strata they lack the strength to act

---

different parts of Serbia, when compared with other republics, three parts of Serbia (Serbia-proper, Kosovo, Vojvodina) are treated separately. The gross national product per capita given illustrates these differences (Yugoslavia=100) Bosnia and Hertzegovina–69, Montenegro–78, Croatia–125, Macedonia–66, Slovenia–203, Serbia Total–90, Serbia Proper–100, Kosovo–28, Vojvodina–118 (SYY 1978:415).

2. 'Semi-professions' refers to occupations requiring secondary level education.

as a group, and, vice-versa, the strata, because of their gender differentiation, cannot become operatively active. Moreover, on the ideological level, women as members of a marginal group exercise a basic ideological function in transmitting the values of the patriarchal culture through childrearing.

The marginal position of women reflects a type of social stratification. The model proposed here implies the existence of two interconnected hierarchies of men and women which together form a third. The span between the social status of men and women increases as one moves down from the higher towards the lower social strata. The smallest distance between men and women is at the highest social status levels, which is also where the likelihood of women achieving that position is smallest.

The lower the stratum to which the individual belongs, the more gender becomes a strong criterion of social differentiation. Gender is more the 'fate' of individuals at the bottom than at the top. From this one can conclude that, in the existing stratification structure, the ascent of women in the social hierarchy 'liberates' them from their gender characteristics, i.e. from the characteristics socially ascribed to them.

This model of the social structure allows the hypothesis that the social promotion of women (in terms of their attaining higher social status) is actually a double promotion, for it liberates them from both the restrictive mechanisms that affect the individual at the lower levels of the social hierarchy and from the restrictions connected with their gender characteristics. Yet, the promotion of women according to this model increases rather than decreases role conflict (profession/family). A woman belonging to the marginal group can overcome her marginality only by climbing the social hierarchy, but the 'price' of her promotion is the tension that exists between the characteristics of her belonging to the marginal group and other characteristics of her social status.

This model suggests an answer to the question of why there are not more women in science. The answer is, that from the bottom to the top of the social pyramid inhibiting mechanisms work upon women, impeding and preventing their promotion, i.e. there is 'social inhibition'.[3]

3. Social inhibition operates through numerous different social institutions: the family, educational institutions, the labour market, professional associations, informal and formal organisation of the enterprises, the mass media, etc.

Social inhibition can be defined as the collection of social mechanisms that impede members of marginal groups from vertical upward mobility. The lower the level of social stratification, the stronger the effect of social inhibition. In order to overcome the 'resistance' of the social structure, members of marginal groups must have additional 'promotional mechanisms' (or such traits as capability, favourable background, support, etc.).

Operating on the level of science are factors related to the position of science in society, the organisation of scientific activities and the characteristics of science itself (its level of development, orientation, dominant paradigms). Society acts on science (as its subsystem) through various institutions, influencing the activities of scientific institutions, the choice of research priorities, the way they are financed and the amount, recruitment policy, promotion in scientific careers, the ideological basis of the theoretical starting points, and the evaluation of the validity and acceptability of scientific views and achievements. (Milic, 1986)

Within scientific institutions, and in the scientific community as a whole, there are specific mechanisms which impede the creative achievements by women and their proper evaluation. Inhibiting mechanisms on the level of science are reflected in the greater difficulties in recruiting women to engage in scientific activities, in the lower evaluation of their creativity,[4] in refuting the right to a specific kind of interest and results (the 'female perspective'), and in the hierarchical and patriarchical system of institutions where scientific work is done. These inhibiting mechanisms become increasingly more pronounced as the area of science becomes more prestigious, closer to the centres of social power and financially more rewarding. Thus, the promotional mechanisms for women to enter these 'more closed' areas of science become more pronounced as such. On the other hand, the concentration of women in certain spheres of science is usually just an extension (only at higher levels) of the gender segregation of occupations.

Since the organisations in which scientific activity is carried out constitute the immediate environment of women scientists, the strongest inhibiting mechanisms are tied in with their struc-

---

4. E.g. more difficult promotion, fewer awards, prizes, fewer important social and political functions, more difficult admission into the Academy of Science, fewer honours etc.

ture. The bureaucratic, hierarchical and patriarchal structure of these organisations as the emanation of the male culture in science, and of traditional, elitist views of science, impedes the establishment of models of communication and work which would be better suited to women. This kind of structure promotes the elitist concept of careers, which is dominated by the importance of the external career (promotion on the basis of rigid, so-called objective criteria, following a straight, continuous line) as opposed to the internal career (increased competence and satisfaction with work).

The bureaucratic formal organisation of scientific activities has its equivalent in its informal organisation. Just as, in their formal organisation, the hierarchies of power and authority are determined by gender so, in their informal organisations, communication is structured on the same principles. Thus, the informal organisation increases the effect of the formal organisation in marginalising women.

Finally, on the third level, the 'individual level', the position of women in science is conditioned by their different life-patterns. The woman scientist's career is complementary to her 'family career'. Furthermore, women scientists promote the quality of life in which there are different valorisations of creativity, a different hierarchy of motives, a different valorisation of power, prestige and money, and a pronounced role conflict. On this level, social inhibition is produced through the tension between the working and the family role. What is important, however, is that the role conflict cannot be resolved by choosing one or the other role (because it is a matter of social compulsion rather than free choice), but rather by changing the relationship between the roles, i.e. by harmonising them.

Lastly, it could be said that the achievement of women in science is marked by tension on all three levels. On the global level this tension will gradually lead to the abolishment of marginality, but also of the system of social promotion itself, which is based on the patriarchal principle. In science, this tension will gradually lead to the re-examination of traditional knowledge, traditional organisations and traditional careers. On the individual level this tension will lead to the re-examination of life models and goals and to the establishment of a discontinuous, multidimensional life model for both men and women.

Women and Higher Education

The post-war development of the educational system in Yugoslavia was marked by strong expansion, embracing women as well, first at lower and then at higher levels of education.[5] For example, the share of women students at institutions of higher education rose from 22.8 per cent in 1938/9 to 45.9 per cent in 1985/6. (Blagojević: 1982; SYY 1987:370)

The highest concentration of all female students is in those faculties where their share of all students ranges from 51 to 75 per cent.[6] These include for example the natural sciences, economy, law, political sciences and philosophy. As many as two-thirds of all female students are enrolled in these faculties. Typically 'feminised' faculties, where women constitute 76 to 100 per cent of the students are the pharmaceutical and pedagogical faculties. (SYY, 1987: 372-3)

The cumulative effects of the varying availability of different fields to women and the inadequate institutional growth of the educational system as such (wherein university faculties and colleges offering education for administrative jobs were over-represented) led to the unfavourable position of women graduates. Between 1945 and 1986, the largest number of women graduated from the faculties of philosophy and philology (46,150); economics (41,780); law (32,470) and medicine (26,468). As many as one third of all women (235,061) graduated from either the faculty of economics or the faculty of law, and only one in every fourteen received her degree from one of the technical faculties. Thus, women are primarily trained to enter office jobs, the medical profession or work as teachers, rather than to qualify for technical careers. (SYY 1987:374, 375)

---

5. Higher education in Yugoslavia includes: university junior colleges (2 years) and universities (4-6 years), and academies of art (4 years). Workers' colleges (2 years) give education for highly-qualified workers occupations (usually technical).
6. In this paper two dimensions of feminisation-masculinisation of educational or scientific fields are analysed: one is the share of women in the total (m+f), and the other is 'concentration' of women in certain fields out of the total number of women. These two dimensions are closely connected, but they are not to be confused.

Women with Post-Graduate Degrees

The decreased share of women with increased level of education continues at the graduate level.[7] In 1986, women accounted for 40.3 per cent of all those receiving a specialist degree in medicine. Also, 31.6 per cent of Master's and 22.6 per cent of doctoral degree recipients were women. (SYY 1987:377)

Even with respect to post-graduate degrees, women were not equally represented or equally concentrated in various fields. Characteristic of specialised studies is the high representation of women in the medical sciences and their high concentration in these sciences. This can be explained on the one hand, by the high feminisation of the intermediary medical schools, and on the other, by the specific position and importance of specialisation in the medical profession. At the Master's level, the highest concentration of women is in the natural and medical sciences, chemical engineering, economics and philosophy faculties. Women account for more than half the Master's degrees at the dentistry, food technology and philological faculties, and for a relatively high percentage (40–50 per cent) of the Master's degrees received in the natural sciences, chemical engineering, philosophy, pharmacology and architecture faculties. In medical schools one out of every two Master's degree recipients was a woman, while women accounted for less than 20 per cent of the Master's degrees at most of the engineering faculties, where female enrolment on the tertiary level is also very low. (S.B. 1988: no. 1690)

The most important indicator of women's inclusion in science is the earning of doctorates. Between the two world wars women accounted for 7.6 per cent of doctoral degree holders. After the Second World War, one out of every five Ph.D.s was a woman. Between 1945 and 1986, 3,311 women out of 16,808 (19.7 per cent) received their Ph.D. Most often, women received doctoral degrees in the natural and medical sciences, and least often in the technical sciences. The most favourable gender structure was in fields which had the highest concentration of

---

7. In Yugoslavia there are 3 types of post-graduate degrees: 1. specialisation (one year long) usually treated as advanced professional training, 2. post-graduate studies (two years long with exams and written thesis), 3. doctoral degrees which require a dissertation (Ph.D.). The last two are scientific degrees.

women Ph.D.s: the natural sciences and medicine. (SYY 1987:377) Even within the scope of feminised fields of study, such as education or the medical profession, it is clearly the rule that as the educational level grows, the share of women decreases. This indirectly shows that the mechanisms preventing the social promotion of women have a stronger (and cumulative) effect at higher levels. Analogous examples are the advancement of women in management, where the higher the level of management the smaller the number of women, or in the professions, where the number of women is much higher on the semi-professional than on the professional level.

The 'Opening' of Science

Women Academics at Universities

In the 1986/7 academic year, there were 6,724 women employed by institutions of higher education, 45.9 per cent of whom were teachers.[8] Women accounted for more than one-quarter of the staff at universities. Although one out of every five university teachers is a woman, women account for 38.6 per cent of the teaching assistants, who will be future teachers themselves. The institutions most open to women teachers are the junior colleges and academies of art; those least open to them are faculties and other university-level organisations. Since junior colleges are at the bottom of the higher education hierarchy, and since art is by tradition the area most open to women, as far as professions are concerned, the fact that faculties, the bulwarks of scientific work in Yugoslavia, are the most closed to women is no coincidence.

The distribution of women university teachers according to their titles shows that 13.3 per cent of all full professors are women (see Table 4.2). The share of women grows as one moves toward the bottom of the hierarchy of positions, which only confirms that the higher social status they hold, the fewer women there are. Since the share of women among teaching assistants is almost twice that of their share among teachers,

---

8. The term 'teacher' at universities in Yugoslavia denotes: full professors, associate professors, docents, senior lecturers and lecturers (except for the last, a doctoral degree is needed). 'Teaching assistants' denotes: instructors (assistants with M.A. degree), language instructors and teaching fellows (without M.A.).

Table 4.1: Women academics in institutions of higher education 1986/7

	Total	Total Women Number	%	Teachers Total	Teachers Women Number	%	Assistants Total	Assistants Women Number	%
Universities	20 554	5 636	27.4	11 684	2 214	18.9	8 870	3 422	38.6
Academies of Art	11 069	291	27.2	842	180	21.4	227	111	48.9
Higher Institutions of Learning	3 369	797	23.7	2 868	623	21.7	501	174	34.7
Total	24 992	6 724	26.9	15 394	3 017	19.6	9 598	3 707	38.6

Source: Statistical Bulletin No. 1683. Beograd: Federal Bureau of Statistics, 1988, p. 9.

Table 4.2: The distribution of female university staff according to position 1986/7

	Total	Women Number	%
Full professors	4 030	538	13.3
Associate professors	2 833	509	18.0
Docents	3 391	817	24.1
Senior lecturers and lecturers	1 949	478	24.5
Teachers – total	12 526	2 394	19.1
Instructors	6 193	2 327	37.6
Language instructors	277	188	67.9
Other assistants	1 004	429	42.7
Teaching fellows and others	1 623	589	36.3
Teaching assistants – total	9 097	3 533	38.8

Source: Statistical Bulletin No. 1683. Beograd: Federal Bureau of Statistics, 1988, p. 10.

since this share is higher in all types of university institutions, and since the trend is towards the share of women increasing as the titles diminish, there can be said to be a trend towards feminisation of university staff, effected primarily through its rejuvenation.

This trend can be observed in all regions of Yugoslavia. With

Table 4.3: Women academics at institutions of higher education, by republics and provinces 1986/7

	Teachers			Teaching Assistants		
		Women			Women	
	Total	Number	%	Total	Number	%
Bosnia and Herzegovina	1 483	260	17.5	1 444	517	35.8
Montenegro	347	37	10.7	135	48	35.6
Croatia	2 677	619	23.1	2 230	925	41.5
Macedonia	973	190	19.5	812	361	44.5
Slovenia	1 533	214	14.0	771	257	33.3
Serbia-Total	5 513	1 074	19.5	3 705	1 425	38.5
Serbia-Proper	3 530	795	22.5	2 289	873	38.1
Kosovo	993	75	7.6	529	157	29.7
Voivodina	990	204	20.6	887	395	44.5
Yugoslavia-Total	12 526	2 394	19.1	9 097	3 533	38.8

Source: Statistical Bulletin No. 1683. Beograd: Federal Bureau of Statistics, 1988, p. 10.

the exception of Slovenia and Macedonia, in the less developed parts of Yugoslavia the share of women among teachers was below the national average, while in the more developed regions it was above average. It is interesting to note, however, that with the introduction of women among teaching assistants, the feminisation of the teaching staff became more pronounced in regions where the share of women among teachers was lower. Hence, inter-regional differences are less pronounced in the share of women among teaching assistants than in the share of women among teachers.

As seen from Table 4.4, women as both teachers and teaching assistants are concentrated in faculties which are more open to women students (social sciences and humanities, medical sciences, natural sciences). The lowest percentage of women teachers is at faculties of machine engineering (6.5 per cent), and the highest at faculties of pharmacology (62.1 per cent). There is a similar situation with teaching assistants. Between these two extremes, the share of women in other faculties is as could be expected: women account for a higher percentage of the teaching staff at faculties of the natural sciences, technology, dentistry and remedial education than in faculties of the technical sciences.

Table 4.4: Women academics (on faculties) according to field 1986/7

	Teachers			Teaching Assistants		
	Total	Women Number	%	Total	Women Number	%
Natural sciences	980	219	22.3	676	300	44.4
Engineering and technology	38 484	517	13.4	2 989	783	26.2
Medical sciences	1 889	507	26.8	2 263	1 027	45.4
Agricultural sciences	1 187	196	16.5	928	338	36.4
Social Sciences and Humanities	3 635	749	20.6	1 996	968	48.5
Total	11 539	2 188	19.0	8 852	3 416	38.6

Source: Statistical Bulletin No. 1683. Beograd: Federal Bureau of Statistics, 1988, pp. 10-23.

The feminisation of teaching assistants has resulted in the fact that in a large number of faculties women account for more than 40 per cent of the teaching assistants: e.g. at faculties of the natural sciences, architecture, technology, medicine, dentistry, pharmacology, veterinary sciences, economics, law, political sciences, philosophy, philology, remedial education, pedagogy, education and music academies, which account for more than half of all the faculties. If this trend continues, with the replacement of the older (largely male) staff and the increasing recruitment of women at lower levels of the university hierarchy, the university teaching staff can be expected to undergo accelerated feminisation. However, instead of treating this change as a chance for the true emancipation of women and their liberation for creative work, it could be viewed as a potential (or already real) danger of marginalising the university as a whole.

Women in Research and Development Organisations

In 1986, 18,502 women (out of 43,240 employees) were employed full-time in research organisations and research divisions of enterprises. More than one quarter of the women are scientists. The proportion of women scientists is lower than that of

Table 4.5: Women in full-time jobs at scientific research organisations (and divisions), by republics and provinces, 1986

	Total	Total Women Number	%	Scientific Researchers Total	Women Number	%	Assistants Total	Women Number	%	Administrative Staff Total	Women Number	%
A	5 231	2 251	43.0	1 732	532	30.7	2 074	930	44.8	630	407	64.6
B	643	221	34.4	191	40	20.9	241	67	27.8	77	49	63.6
C	10 314	5 363	52.0	4 092	1 503	36.7	3 220	1 820	56.5	1 189	878	73.8
D	1 501	585	39.0	604	188	31.1	453	221	48.8	164	98	59.8
E	10 142	4 003	39.5	3 627	939	25.9	4 447	1 758	39.5	992	709	71.5
F-Total	15 409	6 079	39.5	5 400	1 780	33.0	5 060	2 033	40.2	1 834	1 112	60.6
G	13 813	5 488	39.7	4 925	1 650	33.6	4 435	1 751	39.5	1 630	1 014	62.2
H	830	273	32.9	218	39	17.9	344	155	45.1	111	49	44.1
I	766	318	41.5	257	91	35.4	281	127	45.2	93	49	52.7
Total	43 240	18 502	42.8	15 646	4 982	31.8	15 495	6 829	44.1	4 886	3 253	66.6

Source: Statistical Bulletin No. 1668. Beograd: Federal Bureau of Statistics, 1988, pp. 20–27.
A = Bosnia and Herzegovina; B = Montenegro; C = Croatia; D = Macedonia; E = Slovenia; F = Serbia; G = Serbia-Proper; H = Kosovo; I = Vojvodivo.

Table 4.6: Women in full-time jobs at scientific research organisations (and divisions), according to field, 1986

	Total			Scientific Researchers			Assistants			Administrative staff		
		Women			Women			Women			Women	
	Total	Number	%	Total	Number	%	Total	Number	%	Total	Number	%
A	3 591	1 510	42.0	1 870	666	35.6	927	397	42.8	364	264	72.5
B	25 620	9 105	35.5	8 598	2 205	25.6	10 498	3 688	35.1	2 689	1 623	60.4
C	4 815	3 396	70.5	1 448	726	50.1	2 064	1 668	80.8	457	386	84.5
D	4 494	2 007	44.6	1 268	381	30.0	1 061	521	49.1	556	342	61.5
E	4 720	2 484	52.6	2 462	1 004	40.8	945	555	58.7	820	638	77.8
F	43 240	18 502	42.8	15 646	4 982	31.8	15 495	6 829	44.1	4 886	3 253	66.6

Source: Statistical Bulletin No. 1668. Beograd: Federal Bureau of Statistics, 1988, p. 20.
A = Natural; B = Engineering and technology; C = Medicine; D = Agriculture; E = Social and Humanities; F = Total.

assistants, and they account for the highest share of the administrative workers (two thirds). Therefore, the rule in scientific research organisations, too, is that the higher the position, the fewer women there are. Again, the region's level of development is a factor of influence. In the more developed parts (with the exception of Slovenia) there is a higher share of women among scientists, while in the less developed regions, their share is lower. The lowest share of women scientists is in Kosovo Province, the least developed region in the country. Similar to the situation in universities, the fact that the share of women is higher among technical assistants than among scientists indicates a trend towards the feminisation of personnel in research organisations.

The familiar variations in the proportion of women between different scientific fields are also evident in research organisations: women are most often to be found in the fields of medicine and humanities, and least often in the technical sciences. However, the absolute number of women scientists is by far the highest in the technical sciences: 2,205. This means that women scientists in research organisations usually work in environments in which they constitute a glaring minority. This unquestionably reflects upon their professional promotion (and upon their patterns of communication with their male colleagues, as well).

Opportunities and Obstacles

Research on women scientists in Yugoslavia confirmed the basic assumption that they succeed in overcoming the social inhibition that confronts them because of their belonging to a marginal group.[9] They are strongly integrated with the substratum of the creative intelligentsia they belong to. Women scientists define themselves not only in terms of their gender, but above all, by their professions. Still, as members of a marginal group women scientists are not integrated into the social system 'painlessly'. On the contrary, the hallmark of their lives is the extraordinary effort (time, energy) used to overcome social inhibitions in all four relevant spheres: education, science, creativity and family.

9. In this research a mailed questionnaire was used. Around 100 women scientists and artists from Belgrade were included.

A favourable family background turns out to be one of the main resources they tap to overcome these inhibitions. In three-quarters of the cases their fathers are university educated. Both housewife mothers and non-housewife mothers had very supportive attitudes about their daughters' education and professional achievement. The fathers' attitudes were somewhat less supportive. It is interesting, however, that the fathers had a stronger influence on forming their daughters' professional aspiration than the mothers. The fact that fathers had a stronger influence, serving more often as a role model, but offering less support, probably increased the desire of some women scientists for self-affirmation.

Three-quarters of the women scientists chose their profession on the basis of personal interests. In other words, there is a clear intellectual basis for their aspirations. This is also connected with the hierarchy of motives for working in their particular profession. Evaluating their motives for becoming a scientist, women scientists rated creativity highest, followed by independence in life, autonomy in work, usefulness to society and contact with others. In the other half of the hierarchy of motives are the following: money, promotion possibilities, prestige, fame and power. This hierarchy of motives indicates the need to re-evaluate the classical sociological criteria for determining the social status of the individual. At the same time, it points to the predominance of inner over external criteria of success, and intrinsic over extrinsic motivation.

Evaluation of professional experience *vis-à-vis* expectations shows that the expectations of almost one-half of the women scientists were higher than their achievements. Still, more than four-fifths are satisfied with their chosen profession. It is interesting, however, that women scientists do not think they have achieved as much professionally as others believe (family, friends, colleagues). That is to say, their self-assessment of their professional achievements is lower than that of their environment. Behind such responses lies the indirect admission that they could have achieved more in science than they actually did. Asked whether they would have achieved more in their career under some other circumstances, as many as four-fifths of the women scientists replied 'yes'. Most of them cited the problem of adjusting family and professional obligations, but they also frequently mentioned the poor material conditions for work in science (shortage of reading material, lack of sophisticated equipment, travel or stipends) and the unfavourable status of scientists in society (low prestige of scientists). Women scientists feel that work played a key role in their professional success, whereas ability was less important. Only one out of every five women scientists feels that the fact that she is a woman gave her some advantage in her career, but more than two-thirds of the respondents said they had encountered obstacles just because they were women.

This research once again confirms the connection between a woman's family life cycle and her career development. All women scientists who have children consider the childraising period to be the least productive period of their careers. In two-fifths of the cases, women said that family obligations had impeded their professional achievement, while only one-quarter of the respondents felt that family obligations had not slowed down their professional achievement. Asked whether they felt the combination of the roles to be a burden, three-fifths of the women scientists replied in the affirmative. Still, almost all the respondents felt that it is possible to balance a career with family obligations, although it requires a tremendous effort. In keeping with these responses, women scientists listed the following as 'negative' sides of family life: obligations, lack of time and fatigue.

On the other hand, an analysis of responses tied in with the 'positive sides of family life' shows that women scientists derive a high degree of satisfaction from it. Their family life is not a compromise with the environment to prove they are 'normal', but rather a matter of their own personal choice, based on respecting their own emotional needs.

Costs and Rewards

Women scientists are, generally speaking, satisfied with their lives. They are, however, more satisfied with life in general than with their individual (personal) or professional lives. One can therefore conclude that they are satisfied with the combination of the two, but at the same time they are aware that they do not 'extract' the maximum from either their professional or their personal lives. In comparing personal and professional life as a whole, women scientists are least satisfied with their professional life. To them the 'price of emancipation' is both a mental and physical effort to be able to 'function' in two separate spheres.

The qualitative analysis showed that the heavier the burden (more children and less help), the less satisfied women scientists are with both their personal and their professional life. On the other hand, help from the husband, parents, maids, and child care centres not only reduces this burden, but also offers a sense of support which women, like men, need in order to succeed.

There are major differences between women of different generations. The older respondents view family and family obligations as something inevitable, they are more prepared to make compromises and more grateful to their husbands if they are supportive. Younger respondents are more prone to view marriage and parenthood as a choice, and their career and own individual development as an obligation (above all towards themselves). A contributing factor is certainly the fact that younger women, objectively speaking, face fewer obstacles in science because of their gender. The dilemma of a woman contemplating professional achievement in science is no longer 'family – yes, science – ?,' but rather, 'science – yes, family – ?'. Consequently, older women scientists are prone to link their overall success with the family more than with science, whereas for younger women both aspects carry equal weight. It is precisely because they give priority to the family that older women scientists are more prone to have guilt feelings, especially towards their children, which only heightens their sense of role conflict. In short, younger women scientists are much more aware of the 'price' to be paid for abstaining from the double role of career and family, whereas older women scientists are more aware of the 'price' to be paid for combining the two.

The successful adjustment of family and professional obligations is largely related to the distribution of these obligations throughout one's life. Younger and older women scientists have two different models for adjusting their careers and family. Younger women tend first to establish their careers (earning a Master's or doctoral degree), and only then do they interrupt their careers for family obligations. Older women, on the other hand, lean toward the model where they interrupt the establishment of their career to raise children, which produces a greater role conflict and lower lever of satisfaction with professional achievement. This timing is unquestionably connected with the differences in priorities.

Given the severe economic crisis in Yugoslavia,[10] role conflicts, especially among younger women, are greatly aggravated by adverse material and housing conditions, i.e. by limited material resources. However, no less important is the poor organisation of everyday life, which is due not only to the economic crisis but also to long years of insufficient investments in the development of common consumption. The result is that these unfavourable circumstances come to a head in one's late twenties and early thirties: financial insecurity, the problem of housing, the establishment of one's career, marriage, children . . . That is why a strong conflict was evident in some of the younger respondents, coupled with feelings of being powerless to control their own life. Others found a 'solution' by putting off marriage or motherhood.

Still, despite their limited possibilities in terms of material resources, due to the unfavourable social status of the creative (especially younger) intelligentsia in Yugoslav society, women scientists have other resources with which they can partly compensate for the shortage of the former. These are knowledge and the ability of rationally organising the household, wherein (based on a certain hierarchy of needs) the basic needs of the household members will be satisfied. In short, they have sufficient intellectual abilities for developing a maximum rational strategy in an unfavourable social environment.

All the mechanisms of social inhibition which impede women's achievement in their profession, including science, work to heighten the selection process among women themselves. Only those women who are truly exceptional in terms of their abilities, perseverance and energy, and who had the good fortune to develop, under favourable (especially family) circumstances, will be able to reach the heights of science. And because they are exceptional, only they will be able to overcome, often at a great sacrifice, the obstacles that await them.

However, it is important to note that the headway made by women in science will probably lead to the reduction of social inhibition along with changes in different social spheres which

---

10. A good indicator of this crisis is the rate of inflation, 2,500% in 1989.

develop social inhibition in the first place. Briefly, these changes are as follows: changes in *education* – the de-institutionalisation of schooling and permanent education, education for life, not only for work; in *science* – establishing a discontinuous career pattern, de-bureaucratisation and the de-elitisation of creativity and involving a woman's perspective; and changes in the *family* – toward the socialisation of its economic functions and the establishment of co-operative and symmetrical relations between spouses.

Conclusion

In Yugoslavia, as well as in many other countries, certain characteristics distinguish the position of women in science: (1) the percentage of women included in science is lower than that of men; (2) the lower the position in science, the more women there are; (3) the availability of different fields to women depends on the position of the sciences: the more prestigious the science the more closed it is to women; and (4) the most open sciences to women are those that men have already 'left', or those that are the most connected with women's traditional roles.

It seems that improving the position of women in science is inseparable from changing present models of organising scientific activities. Non-hierarchical team work, co-operation as opposed to competition, the division of authority and responsibility, self-organisation as opposed to external organisation, flexible rules as opposed to rigid organisational rules, and a multi-dimensional and discontinuous career are better suited to women and to new professions in general. This kind of reorganisation of scientific activity would also imply that the valorisation of achievements by the individual in science would be the result of the individual's genuine contribution rather than his/her place in the hierarchy.

References

BLAGOJEVIĆ, M., 'Neke tendencije razvoja obrazovnog sistema u Jugoslaviji' (Some Tendencies in the Development of Educational System in Yugoslavia), *Mlada generacija danas* (Young Generation Today), Beograd: NIRO 'Mladost', 1982.

——, 'Drustveni polozaj zena strucnjaka u Jugoslaviji' (Social Status of Women Professionals in Yugoslavia), unpublished Ph.D. thesis.

MILIC, V., 'Naucni potencijal SFR Jugoslavije: rast i problemi' (Scientific Potential of Yugoslavia: Growth and Problems), *Zbornik Filozofskog fakulteta*, No. 14, Beograd, 1986.

SB-*Statisticki Bilteni* (Statistical Bulletins), Savezni zavod za statistiku, Beograd.

SYY-*Statisticki godisnjak Jugoslav* (Statistical Yearbook of Yugoslavia), Savezni zavod za statistiku. Beograd, S 25.

5

Women and Science in Bulgaria: The Long Hurdle-Race

Nora Ananieva

Introduction

In the course of a radical reform of the economic and political system of Bulgaria, the women's question is placed in a new historical and theoretical context. To follow and analyse the chain linking women with science and socialism will imply full compliance with the requirements of the Marxist dialectical method, i.e. that social phenomena should not be perceived 'as a complex of complete things, but as a complex of processes'. (Marx and Engels, vol. 21:299) For the founder of the Bulgarian Social Democratic Party the question of the emancipation of women becomes one with the question of the economic emancipation of mankind. (Blagoev and Dimitrov, 1979:14)

The present situation displays the 'boomerang' effect of the previous self-satisfaction: surveys of women's problems are still in their rudimentary stage. In women's studies, the status of women in science occupies but a marginal place: figures are insufficient even for the purpose of analysing general trends, mainly because of the differences in the indices and the criteria applied. A unique contribution so far is the work on women's access to science and the factors counteracting the discrimination against women. (Ananieva, 1984)

The point of departure for this survey is the peculiarities of the political history and the history of science in Bulgaria. Such an approach overcomes the vulgar determinism, the dogma of the socio-economic and political changes automatically reflected in public consciousness and psychology. The steadfastness of some ideas and stereotypes in consciousness, their 'transsystematic' nature are demonstrated convincingly when penetrating the real situation in which women scholars live and develop.

The vicissitudes in the historical fate of the Bulgarian nation and the Bulgarian women are preserved in the public consciousness and social psychology of the people. As the lines on a human hand can tell, so can they tell both about the traces of the five-century long yoke and the gleams of the ineradicable national self-awareness, the patriarchal way of life and 'the opening' to the world, the revolutionary leap into a radically changed social reality and the inevitable post-revolutionary sobering, the awakened expectations and the fading out of the illusions.

The Historical Background or the Favourable Impact of a Discriminatory Stereotype

Though the beginnings of the Bulgarian state date back to 681, its spiritual present is deeply rooted in the time of the Renaissance. This was the epoch which united the impulse for enlightenment with the idea of national liberation, the strong national cultural roots with the influence of the ideas of the great French Revolution. That epoch found the Bulgarian woman confined to the family, subservient to her husband and entirely preoccupied with the care of the household and children. A host of factors determined that situation. On the one hand, the low level of education in society, the ascetic Christian views, the Ottoman yoke and the general influence of the Orient (Pundeva, 1940:9) accounted for this status. However, on the other hand, the Renaissance also found a smouldering but inextinguished Slavo-Bulgarian tradition. In 'Old Bulgaria' the woman was an all but equal member of the family and the patriarchal community, her opinion was respected in resolving questions pertaining to property and other family and lineal issues. That tradition offered favourable support to the ideas of the early Bulgarian enlighteners. In 1840 the first girls' school was established. Accordingly, in the last decades before the liberation such schools were opened in almost all big towns in the country. Enlightened and rich Bulgarians subsidised the training of women teachers in Russia, Serbia and Greece. At this time intellectuals passionately argued that it was impossible to enlighten a nation where men and women were not equally educated. (Pundeva, 1940:42)

Situated on the borderline between Europe and Asia, Bulgaria has long been an arena of sharp confrontations for different civilisations. Such confrontations, though in different forms, did not cease after the liberation which was in fact a bourgeois democratic revolution. The Tirnovo Constitution, though markedly advanced for its time, did not establish even the legal prerequisites for the emancipation of women. Therefore, the whole period of capitalist development was characterised by the struggle for equality and social liberation of Bulgarian women.

The statistics from the first population census (1946) after the Socialist Revolution gives a general idea of the status of Bulgarian women inherited from the pre-revolutionary times: 79.2 per cent of all women of economically active age were engaged in the labour force: women constituted 45 per cent of the total economically active population. Yet, the Revolution inherited a backward agrarian country; thus the greater part of the economically active female population was concentrated in agriculture. (Dinkova, 1980:16–18) The employment structure was such that the involvement of women was totally negligible. Only 1.3 per cent of the economically active women in 1946 were specialists with higher and secondary school vocational training. Furthermore, even those women who were employed outside the agrarian sector were mostly occupied in the auxiliary and lower-paid positions. The legal system endorsed a secondary position for women in society; it deprived them of electoral rights and of the right to decide on their own on the family possessions.

From this perspective, the changes which took place after the 1944 Socialist Revolution were radical and complex indeed: they were socio-economic, political, legal, and socio-psychological. The statistics from the 1975 census showed an almost even distribution of men and women in the different strata of the social structure. Some lagging behind was still observed in the case of the employees-specialists with higher education. However, the 39.5 per cent employment rate attained by women is the most striking expression of the revolutionary change.[1] In 1946, out of all specialists with higher education, women accounted for about 60 per cent of the school teachers. This ratio

---

1. At the beginning of the century only 13.9% of the female population were literate. The first women students were admitted to higher education in the year 1901.

reflected both the professional structure of Bulgarian specialists from the pre-revolutionary period, and the influence of the traditional stereotype of women's professions. Ever since the establishment of the first girls' schools, society had perceived the educated woman as a teacher. In fact, it was along the lines of that stereotype, paradoxical as it was, that the first breakthrough of women in science took place. Out of the total number of lecturers at the higher education institutions in 1946, women amounted to 3.4 per cent. (Dimitrov, 1974)

The road leading Bulgarian women from teaching in secondary schools to higher education usually passed through training at prestigious universities abroad, which is where their scientific interests were shaped. Women's entry into the higher education system was also aided at its initial stage by the fact that the first higher education institution in Bulgaria, i.e Sofia University, was established in 1888 as the Higher Pedagogical School (*History of the Sofia University*, 1988:5).

Since the leading figures of the National Renaissance were also educational functionaries, this profession had high prestige in society. The association of educated women of the teaching profession reflected this prestige.

Typical of the Renaissance tradition was also the fact that when in 1869 the first scientific institution was initiated and established, it was called 'The Bulgarian Literary Society'. Later this institution was turned into 'The Bulgarian Academy of Sciences', which originally had three branches: history-philology, natural science-medicine and 'state-science'.[2] (*Bulgarian Academy of Sciences*, 1958:7)

The historical development of these branches determined the initial history–philology orientation of women who devoted themselves to a scientific career. For instance in the *Manual of Bulgarian Historians* out of a total of 751 names, women amount to 133, i.e. about 18 per cent. (Holov, 1981) Though most of these names are contemporary, the place of Bulgarian women in history and philology dates further back to the past.

The historico-philological orientation gave rise to an interest in other fields of social sciences. The first women who won recognition in social sciences were graduates from Sofia Univer-

---

2. Equivalent to political science, which has not been accepted as a discipline until recently.

sity's faculties of Slavonic, Classical and Western philologies in the 1930s and 1940s. Since Bulgarian women played a traditional and decisive role in the formation of the spiritual and material culture of the Bulgarian nation, their participation in these scientific fields was natural. The distribution of academic researchers in social sciences throughout the 100-year-old history of the Bulgarian Academy of Sciences reflects this situation. Out of a total of 212 names, 43 are women (20 per cent). These women are most strongly represented in archaeology (33 per cent), library science (60 per cent), linguistics (about 50 per cent), ethnography (almost 40 per cent) etc. (*Hundredth Anniversary of the Bulgarian Academy of Sciences*, 1972). Furthermore, as mentioned earlier, today as in the past, there is a marked interest shown by women researchers in the study of national origins. These include research in fields such as mediaeval art, mediaeval and early Slavic archaeology, Bulgarian cultural history, Bulgarian Renaissance and Bulgarian association with Western Europe in the eighteenth and nineteenth centuries, mediaeval Balkan and Bulgarian history, Byzantology, etc.

The stereotype of women teachers and women researchers in the humanities gradually paved the way for women in the other fields of science. In 1922/3 at Sofia University there were eight women out of a total of 132 lecturers. However, they helped to overcome the traditional barriers. Although they were still considered to be rare 'exotic flowers', women were no longer total strangers in the university. Breakthroughs in other areas seemed more likely to take place against this background. And they were quick to come. Sofia University welcomed its first woman nuclear physicist, first as an associate professor and later as a professor, in the 1930s. This woman of high qualification was the head of the Department of Nuclear Physics at Sofia University for ten years. She was also the chief of the Laboratory on Radioactivity and Department head at the Physics Institute of the Bulgarian Academy of Sciences (*Hundredth Anniversary of the Bulgarian Academy of Sciences*, 1972:92). Similar breakthroughs were observed in other spheres of science, too.[3]

---

3. The first and the only female full member of the Bulgarian Academy of Sciences in its 100-year long history was a specialist in genetics (*Hundredth Anniversary of the Bulgarian Academy of Sciences*, 1969).

The Heavy Lot of the Dialectical Law: About the Transition from Quantity to Quality

According to the Unesco statistics on women's participation in R & D activities, Bulgaria has a prominent place with 40 per cent of the R & D personnel being women. Furthermore, the number of women shows a consistent annual mean increase of about 2.4 per cent. (Unesco, 1980:6, 8, 11, 32). On the basis of these figures there is reason to speak of a real 'invasion' of women in science. However, the analysis accompanying these data is insufficient, at least in two aspects: firstly, it contains the distribution of women by branches of science and not according to the places they occupy in the scientific hierarchy, which fails to assess the opportunities for promotion; secondly, this analysis over-emphasises the role of the planned economy and neglects the entire range of factors having an impact on this process. Without an analysis of these factors, women's entry into science in larger numbers appears to be simply a reflection and consequence of the rate of emancipation in other social spheres (employment, education, qualifications, social life), but this does not take into account the role of women in science.

Data for the period after the published surveys carried out by Unesco show a consistently high involvement of women in scientific research activities. The percentage of women increased from 34.21 per cent in 1978 to 37.93 per cent by 1987. The continuing increase in the ratio of women researchers has been preserved against the background of the relatively high rate of increase in the scientific potential of the country, which is a result of the scientific and technical modernisation. Given this tendency, however, two things should be mentioned. Firstly, despite the overall tendency of an increasing participation of women, the ratio of women in engineering sciences is relatively low. Secondly, a noticeable 'leap' in the participation of women in agricultural sciences is observed as their ratio has increased from 11.46 per cent to 28.22 per cent (see Figure 5.1).

The first case further contrasts with the high ratio of women engineers. The situation can be explained both by the pragmatic professional orientation of women engineers and the influence of the stereotypical image of science as a male profession. Also, women's interest in the engineering field has begun to abate. This is both because of the relatively low pay and the undesir-

Figure 5.1: Distribution of women scholars by sciences and main subjects (comparative data in percentages)

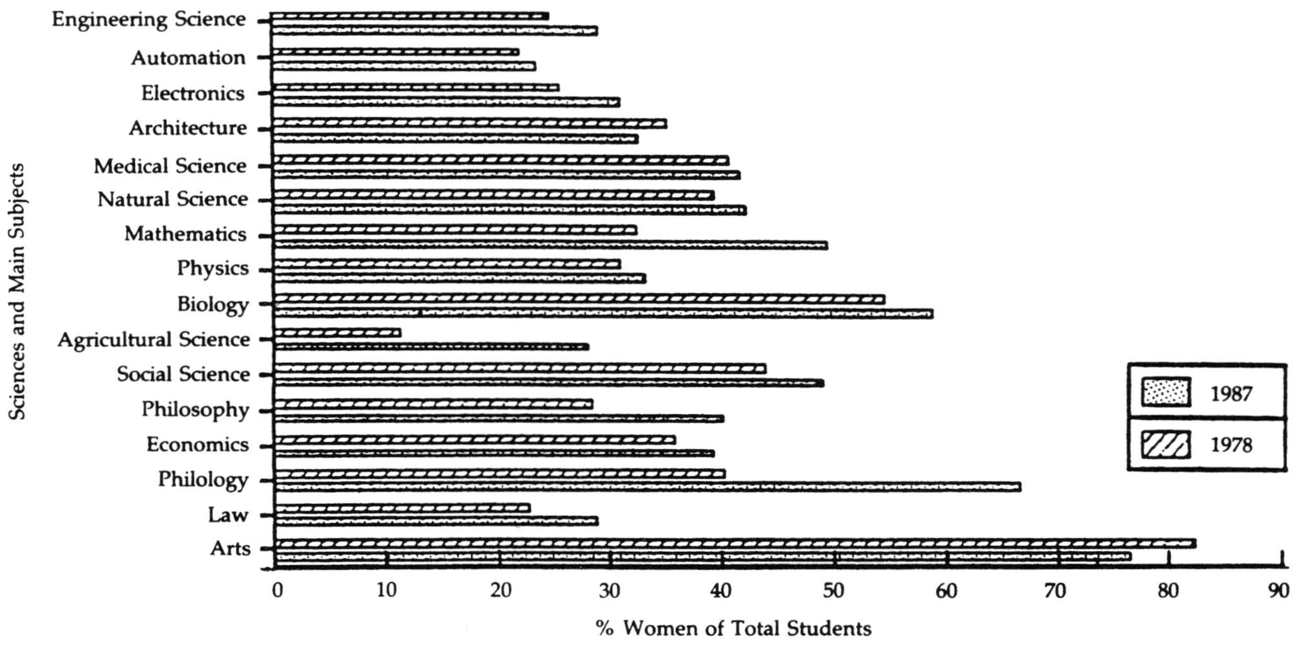

Source: Bulgarian Statistical Yearbook 1978–1987

able characteristics of the profession. What is more, under the present conditions there is no longer any need to 'compensate' in areas that were once closed to women.

Agricultural sciences, on the other hand, have never been of special attraction for women, despite the fact that the first woman member of the Academy of Sciences came from this background. It is possible that this attitude is deeply rooted in the former 'agrarian way of life' and in the continuing aspiration to get rid of it. The stereotype of the woman teacher was rendered concrete in the high percentage of women academics at universities and higher institutes of learning. Within a period of ten years there was a decrease due to the accelerated development of applied sciences directly related to the demands of the economy. Quite a few women with scientific interests preferred to be employed in R & D units of enterprises, where they are both better paid and able to achieve a speedier materialisation of their scientific results in practice (see figure 5.2).

Of great interest are the data on the distribution of women scholars by degree and academic position. They can be illustrated in the form of a pyramid with a broad foundation and a thin needle-shaped peak. A tendency for an increase at the higher level, which means an increase, though at a very slow rate, in the number of full professors, can also be observed. Of special significance for promotion in the scientific sector in Bulgaria is the advanced doctoral degree, i.e. D.Sc., which is almost an absolute prerequisite for a professorship and is conferred on the basis of a significant contribution made to science. One of the serious barriers to the advancement of women in science is the system requiring advanced doctoral theses to be defended before specialised scientific councils (almost all of which consist only of men) (see Figure 5.3).

The figures for women doctoral students and those with Ph.D. and D.Sc. degrees show a continuing increase, which in the case of certain science disciplines exceeds half of the total number. Compared to that, 'the zeros' in the engineering sciences and in the natural sciences are very striking (in 1978). Growth is observed in all spheres (except in the sphere of engineering sciences): the highest increase (from 15.06 to 25.28 per cent) is in the field of medical and social sciences (see Figure 5.4 and Figure 5.5).

Against a background of factors favourable to the access of

Figure 5.2: Women academics in universities and higher institutes (comparative data in percentages 1978–87)

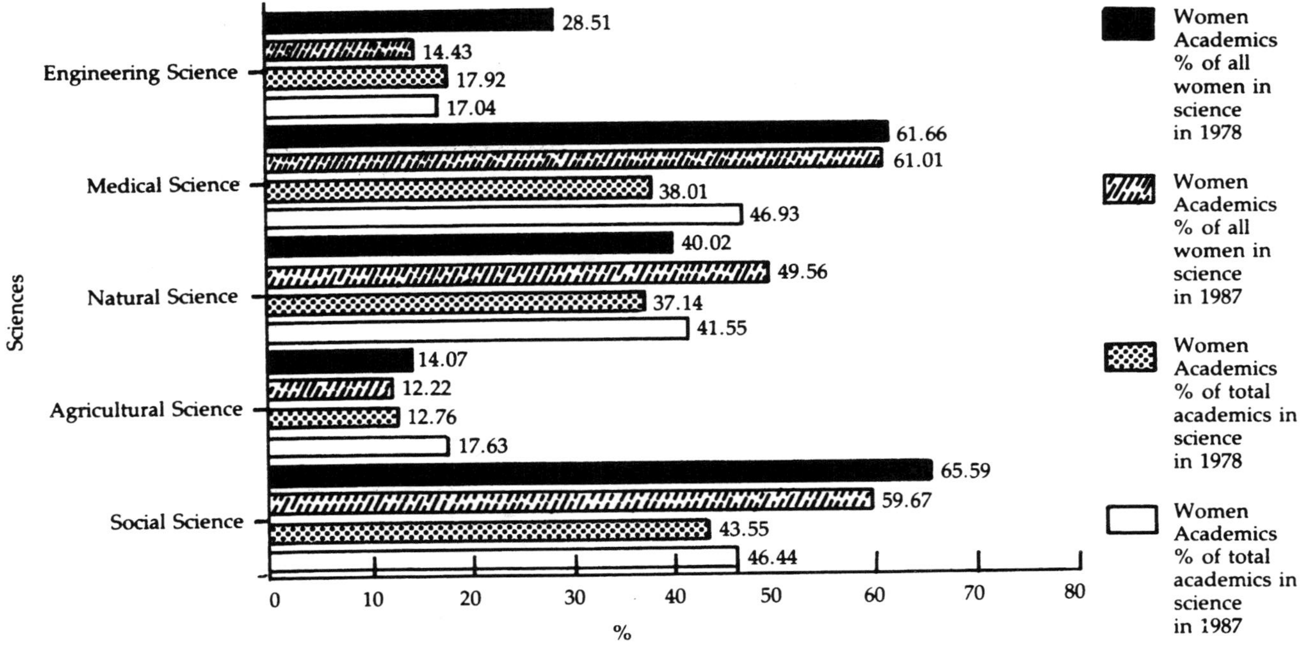

■ Women Academics % of all women in science in 1978
▨ Women Academics % of all women in science in 1987
▦ Women Academics % of total academics in science in 1978
☐ Women Academics % of total academics in science in 1987

Sciences:
- Engineering Science: 28.51, 14.43, 17.92, 17.04
- Medical Science: 61.66, 61.01, 38.01, 46.93
- Natural Science: 40.02, 49.56, 37.14, 41.55
- Agricultural Science: 14.07, 12.22, 12.76, 17.63
- Social Science: 65.59, 59.67, 43.55, 46.44

Source: Bulgarian Statistical Yearbook 1978–1987

Figure 5.3: Women scholars by degree and academic position

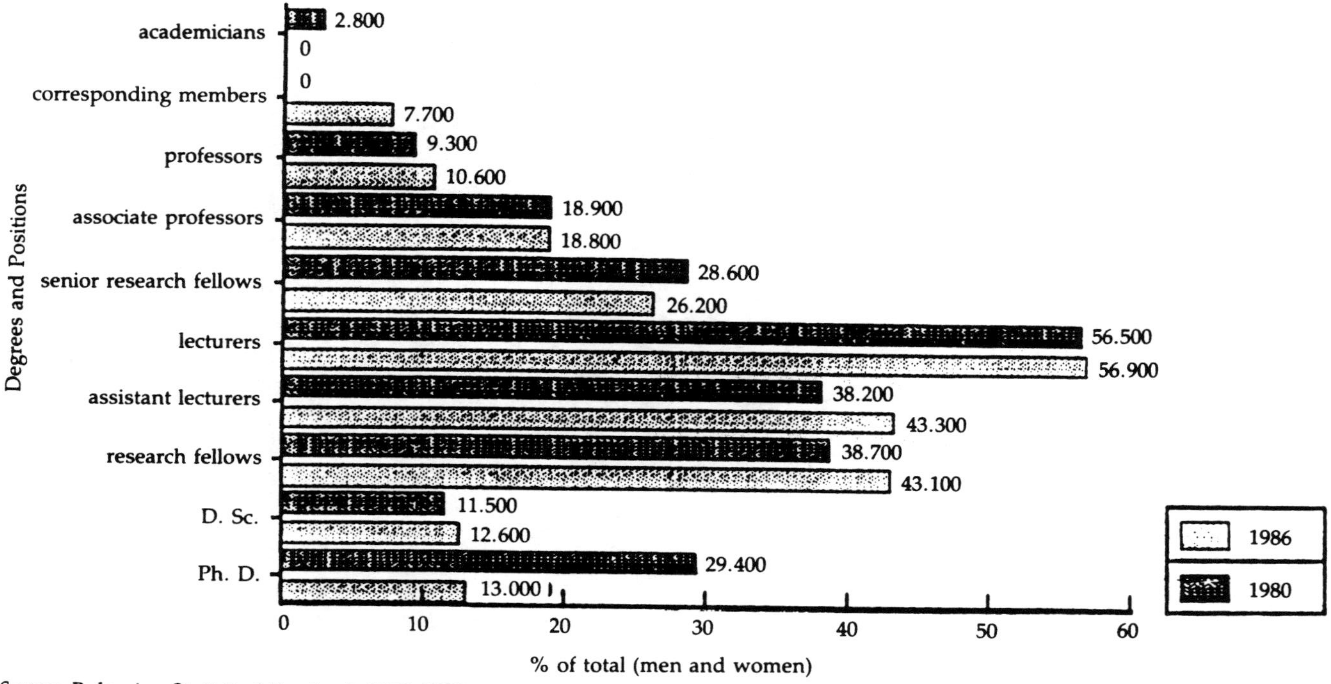

Degrees and Positions	1980	1986
academicians	2.800	0
corresponding members	0	7.700
professors	9.300	10.600
associate professors	18.900	18.800
senior research fellows	28.600	26.200
lecturers	56.500	56.900
assistant lecturers	38.200	43.300
research fellows	38.700	43.100
D. Sc.	11.500	12.600
Ph. D.	29.400	13.000

% of total (men and women)

Source: Bulgarian Statistical Yearbook 1980–1987

Figure 5.4: Women doctoral students and women with higher degrees received from Bulgaria and abroad (percentages of total men and women)

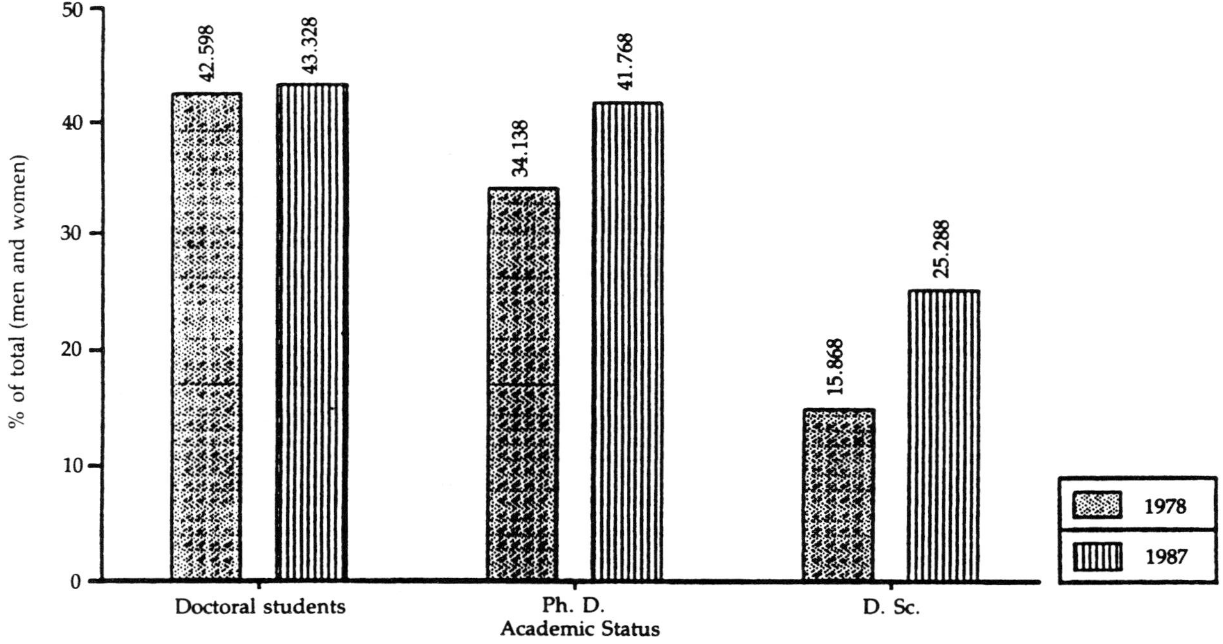

Category	1978	1987
Doctoral students	42.598	43.328
Ph. D.	34.138	41.768
D. Sc.	15.868	25.288

Source: Bulgarian Statistical Yearbook 1978–1987

Figure 5.5: Women doctoral students and women with higher degrees received from Bulgaria and abroad (by sciences, percentage of total men and women)

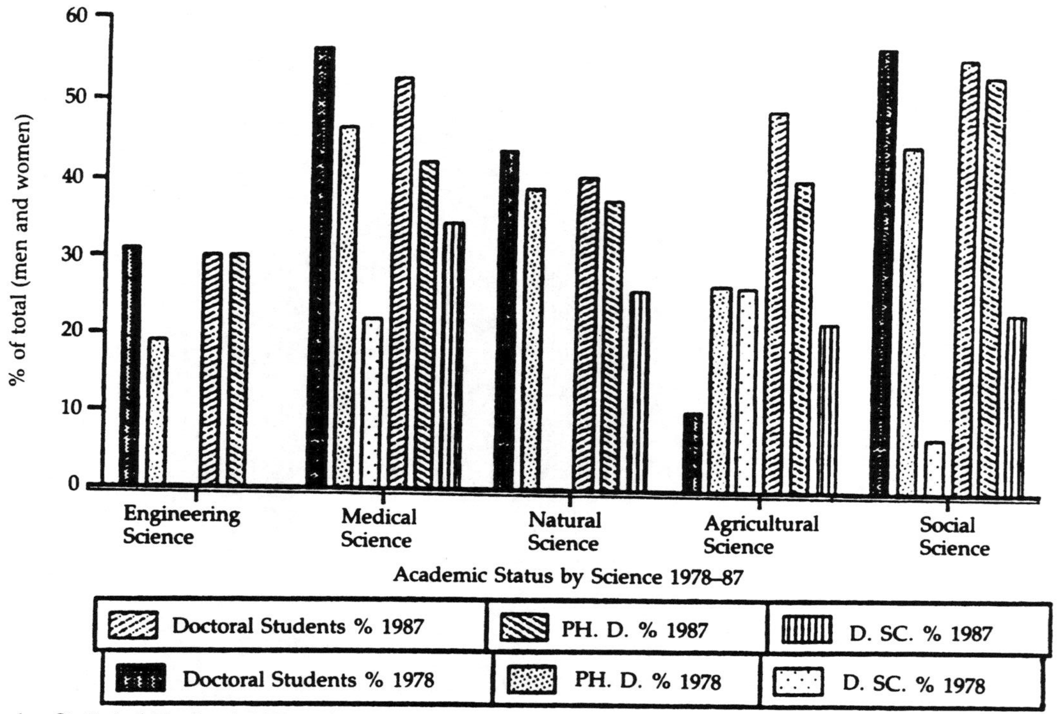

Source: Bulgarian Statistical Yearbook 1978–1987

women to science which has existed for more than 45 years – since the Socialist Revolution – there remains the overall delay in the promotion of women within the scientific system. A century-long development of the natural sciences within the system of the Bulgarian Academy of Sciences, up to 1971, witnessed only five women professors. In the same period, in the field of social sciences, out of a total of forty-three women with academic ranks, there was only one woman professor. In the following period about one-third of them reached the rank of professor. Between 1945 and 1970, the Medical Academy in the town of Plovdiv employed a total of 538 university teachers, out of whom only 105 were women; of these only four were associate professors and one was a professor. (Medical Academy, 1971)

Another indicator of the nature of women's careers is provided by their participation in the governing bodies of scientific organisations. Only scientists with academic ranks have access to the two leading institutions – the Scientific and the Directorate Councils – which a priori narrows the basis for the participation of women.

In the second quarter of 1989, elections were held for the leading bodies of these institutions in the entire system of the Bulgarian Academy of Sciences. They were carried out with the participation of all scientists, and every unit, organisation or individual could nominate candidates. Voting was by secret ballot in several rounds. The outcome from the elections showed a relative increase in the number of women in the governing bodies. In sixty-six research institutions (or laboratories) within the system of the Bulgarian Academy of Sciences, three women (or 4.4 per cent) were elected directors. This outcome meant preservation of the same percentage as in previous times, with only a change in the persons elected; also nine women (or 14 per cent) were elected deputy directors. Women's highest representation was among the newly elected women scientific secretaries, amounting to 21 per cent. This position, however, has a dual connotation. While, on the one hand, it is a reflection of scientific recognition, on the other hand, it complies with the stereotypical image of women as 'organisers of domestic work'. All in all, women account for 13 per cent among the leaders of directorate bodies.[4]

4. Unpublished data were collected and calculated personally by the author.

The elections showed that under a democratic procedure women can achieve leading positions in the research institutes with the strongest concentration of women, in such branches as: genetics, physiology, microbiology, organic chemistry, history and theory of urban planning, ethnography, archaeology, Bulgarian language, history, sociology, economics, etc.. At the same time, it is interesting that in forty-one research institutions no woman was elected to a position of leadership, even in institutions where women have the decisive voice in the electoral bodies. In these cases a concrete analysis of the inner factors, including the socio-psychological factors which influence the behaviour of participants in elections, is needed.

A Macro–Micro Linkage in the Analysis of the Factors Influencing Women's Advancement in Science (A Case-Study)

The Institute for Modern Social Theories at the Presidium of the Bulgarian Academy of Sciences is a complex scientific research institute which studies the latest tendencies and schools in modern social thought, as well as the current political processes in East-West relations and in the development of different political systems. Comprising scholars of various research orientations (philosophy, economics, law, politics, sociology, pedagogy, etc.), it investigates key political and sociological themes on an inter-disciplinary basis. Therefore, the place and role of women in its activities could be considered as representative of women's position in social sciences.

Out of a total of sixty-seven scientists employed in this Institute, twenty-eight are women (or 42 per cent). In terms of degrees and ranks the picture seems to be typical: women account for 72 per cent of all research fellows (eighteen women out of a total of twenty-five persons), and for 37 per cent of all senior research fellows (or nine women out of a total of twenty-seven). There is only one woman professor (6 per cent) out of a total of fifteen professors.

This ratio favourably influenced the representation of women on the newly elected leading bodies of the Institute. For the first time in the history of the Institute a woman (the only woman professor) was elected with the highest percentage of votes cast by secret ballot, which implies the support of both men and women. Women's representation on the Scientific Council has doubled from 5.4 per cent to 10.8 per cent (or from two to four members out of a total of thirty-seven members). The inquiry which was carried out[5] helps the clarification of certain tendencies. It is divided into two groups: **(A)** all women members of the Scientific Council;[6] **(B)** all women scholars (twenty-one out of a total of twenty-eight women) who were present at the time of the inquiry.

(A) The age distribution of women members of the Scientific Council is as follows: one woman under fifty years, two women over fifty years, and one woman of sixty years. Their average age is about fifty-four years.

Two of the women were elected members of the Scientific Council for the first time, and two of them for the second time. In three of the cases at least one of the parents had higher education, in one case the parents were workers, and in another case the parents were very poor. Only in one case had the childhood been spent in Sofia and in all other cases outside Sofia, including in a village. Nevertheless the stimuli for schooling had come above all from the family, and mainly from the fathers. The family atmosphere encouraged an enrichment of knowledge, although the principle of equality of the parents in the family was noted rather as the exception.

Taking into consideration the fact that all these women had grown up in a complicated historical period,[7] it is obvious that contradictory factors influenced their orientation towards science. Despite their good higher education and despite their early scientific interests, their way to science was not a direct one. All of them started work early, some of them won recognition in other professions such as journalism and public administration before they embarked on scientific work, and early in their lives they were taught to combine many responsible activi-

---

5. The inquiry encompassed a sample of all women scientists in the Institute for Modern Social Theories.
6. It does not involve only permanently appointed scholars from the scientific staff of the Institute.
7. The oldest of them took an active part in the struggle against fascism, was sentenced to death by the fascist court but escaped death because she was a minor.

ties. All four women are married (one for the second time), three have one child each, and one has two children. All of them assess their family duties and caring for their children as hampering their scientific careers. Despite the help rendered to them by their parents, the heavy burden lies on the shoulders of women. This is also reflected in their time budgets. All of them think that they pay a heavy price of deprivation of rest, night work, limited cultural activities, limited contacts with friends, etc.

The scientific careers of the four members of the Scientific Council have been rather contradictory. The defence of the doctoral theses was not immediately after graduating. Only two of them had ventured to stand for defence of their D.Sc. degree. However, the scientific achievements of all of them are very convincing: they have produced a great number of publications (some of them abroad), as well as articles in scientific journals. They are heads of research teams and are participants in international scientific activities. All of them speak Russian and English, some also speak French, Spanish, Italian and German.

Two assessments are of special interest: those on the status of women in society and in science, and those on the factors hampering women's careers. The opinion is prevailing that in our country women are not objectively unequal, but, for many reasons, the conditions for career promotion are very complicated. Two clear-cut factors hamper their career promotion. Firstly, the various duties of women with regard to household, everyday life and the family, and the painful, though not impossible, combination of these activities with a scientific career. Secondly, the continuing scepticism regarding women's scientific abilities. This double, objective–subjective hindrance creates a heavy load. It calls for super-organisational abilities, strong will and ambition in order to succeed. Nevertheless, there is predominant satisfaction with the chosen scientific orientation. Proposals for change refer to the development of the material and spiritual culture of society, the development of higher culture, ethics, tolerance and intelligence among male scholars, etc.[8]

---

8. In order to compare the results, one more woman was interviewed – a university professor, department head, former Dean of Faculty of Philosophy at the Sofia University. The case shows a career development where the doctoral studies immediately follow higher education, the scientific orientation is initial and final, the positions of recognition are won considerably early. The impulses

The inquiry proves that women who, to varying extents, earlier or later, have achieved different alternatives of scientific career promotion, show satisfaction with the attainment of the 'forbidden fruit'. Simultaneously, there is a bitterness due to the price they had to pay: the lost time in tiresome conflicts for self-realisation, the humiliation suffered because of neglect and disregard, the failed opportunities for a better personal life.

(B) The inquiry on the staff of women scholars at the Institute of Modern Social Theories reveals a number of common, but also some specific phenomena and characteristics. Out of a group of women with an average age of forty-two years, only two are single (thirty-one and thirty-two years of age). All the others have gone through 'the school of marriage' and all have children (one of them has three children, nine of them have two children each and the other nine have one child each). Eleven of them are currently married and eight of them are divorced, which is a very high average for the country.

These women scholars have a heterogeneous social background. Taking into consideration the fact that the majority of the staff is up to forty years of age, i.e. their parents have also got opportunities for education, it is natural that in the families of fifteen of the women there are university graduates, four women come from workers' families with parents of primary or secondary school education, and two women come from a village. In seven of the cases some of the parents are also engaged in scientific work. However, the majority underscore the strong influence of the father while admitting that in most of the families the principle of equality has not been fully realised. This implies that even when preserving the patriarchal way of thinking, fathers are apt to defend equality and emancipation for their daughters.

With a view to their education, qualification and foreign language training, women scholars at the Institute for Modern

---

come from an intelligent family including other scholars, as well. The husband is also a professor. The principle of equality is something natural both in the family of the parents and in their own. From this, to a great extent, 'cloudless career', it is logical to draw the conclusion that the behaviour towards women 'depends on women themselves'. However, it is stated that women are faced by greater difficulties, as the difficult combination of the various activities is stressed as the greatest hampering factor.

Social Theories have fully utilised the opportunities created in the country. Fifteen of them are Sofia University graduates,[9] four women are graduates from educational institutions in the USSR, one graduated from the Higher School of Engineering and one from the Higher Economics Institute. All of them have completed their doctoral studies; they have obtained their Ph.D.s. All speak Russian well and at least one Western European language. They make use of some other languages and some of them are also good at rare foreign languages, such as Chinese, Japanese, Arabic and Greek. A high percentage have made their specialisations in different countries: the USSR, USA, United Kingdom, Greece, the People's Republic of China, the GDR, Japan, etc.

Most of the doctoral studies coincide with marriage and the birth of children. There are many difficulties at the initial stage of a woman's career. Almost all young families have financial difficulties, and as a rule they do not have normal housing conditions. Therefore, it can be expected that at least some of them give up. However, there is enough ambition to surmount the difficulties. The patriarchal tradition plays a positive role: child care is often entrusted to grandmothers. It is not by chance that the new social law gives grandmothers the opportunity to take paid leave for the care of grandchildren (an opportunity for a family to make its choice). At the initial stage of marriage, young fathers are more willing to share the burdens of the household.

The problems most often start occurring after the defence of the thesis. All the respondents are faced by difficulties in combining family with promotion. A rather high ratio of the time budget is devoted to household and family, which leaves hardly any time for creative work. New problems start accumulating gradually: a difficult breakthrough in the publishing houses, lack of confidence in the research teams comprising, above all, 'successful' men, limited opportunities for rest and private life, growing nervous tension, etc. It is in this period of time that a large part of women get divorced. There is a decrease in the opportunities of the parents to render assistance and their own financial conditions do not allow help from outside. 'I have

---

9. Three in law, four in journalism, four in different kinds of philology, two in pedagogy and two in philosophy.

almost no help any more, and this is my hardest struggle in life', writes a young woman who has just attained her academic rank and thus has just succeeded in making a breakthrough.

The comparison of the age groups of men and women scholars with academic rank shows a definite delay of women's promotion. There is a forcible pause associated with a kind of crisis. Their self-confidence is different from that at the beginning of their career, though it has not yet disappeared completely. In their self-evaluation few women (only two) dare to speak about their creative talents, most of them place a stress on theoretical thinking, on perseverance, on self-discipline. Caution is predominant in the formulation of their objectives: 'to write an interesting book on a significant topic', 'to write in a non-traditional way', etc. Only one of the women explicitly formulated her intention to defend a D.Sc. thesis.

The greater part of the women view positively the behaviour of their immediate chiefs, both men and women, as well as the atmosphere for the promotion of young people in the Institute itself. This leads to the conclusion that the unsolved problems of living conditions, supply, services and culture in human relations have their effects on the self-confidence and mentality of women scholars. Therefore, in the main, there are two proposals for a change in the situation: the attainment of real equality between the two sexes, and measures to ease the daily lives of married women. Quite a large number of women state their dissatisfaction with their insufficient involvement both in research and civic activities, in editorial boards and in scientific forums. Therefore, there is a real hunger for creative expression, but also for positions with additional responsibility. Obviously what is meant here, is 'the desired burden'. It is not by chance that the women scientists with most responsibilities are also the most productive. In their case the self-discipline grows into super-organisation, and their creative activities are a stimulus to additional energy. However, sixteen women out of a total of twenty-one respondents state that they have no other position. Few women are involved in research teams. This prevents them from assessing their own performance compared to their colleagues.

Also of significance is the role of women scholars who have already won recognition and got positions. The inquiry shows the positive fact that these women help their women colleagues.

Some of the respondents state clearly that they have taken a woman scientist as a model. But the phenomena revealed through the inquiry point to the need for still more different, efficient and, in some cases, even an emergency form of mutual assistance.

Maybe the hardest burden for women scholars is their lack of confidence. Estimates of the behaviour towards women vary from: 'It depends on the women', to 'the inevitable discrimination'. Without carrying things too far, according to the prevailing opinion there is still disregard, suspicion, and indirectly women do not enjoy equal rights with men. The traditional stereotypes happen to be extremely stable and the continuing struggle against them is an additional psychological burden for women. However, experience shows that these stereotypes can be overcome in the immediate creative work. At the more advanced stage of career promotion in science they even cease to be felt as such.

This complex situation, especially in the case of young women in science, could bring about a feeling of resignation. Such remarks could also be found in this inquiry: 'The situation cannot be changed', 'there is no sense in the proposals', etc. These comments are not predominant though. Prevalent is the assessment that there are real opportunities for a woman to win recognition in the scientific field if she has got the necessary abilities and ambition. The subjective evaluation of 'reliance on one's own strength' is a kind of expression of self-confidence that has not yet disappeared despite all difficulties. 'I could have a more peaceful life', writes one of the respondents, 'but still I feel that work will attract me, the work associated with the "grey matter".' 'I am pleased', says a woman, 'because I have reached everything myself and in spite of all difficulties'.

And yet, both the present inquiry and the general analysis of the place and role of Bulgarian women in science bear witness to the fact that women's promotion and realisation of full value depends on a complex set of factors. To identify and analyse these factors is the precondition needed for the elaboration of a platform for the necessary changes.

Conclusion

The situation in the People's Republic of Bulgaria, forty-five years after the Socialist Revolution, enables us to draw the conclusion that, in the main, the problem of equal access for women to science has been solved. The objective requirements of socio-economic development, including science strategy as well as political and ideological factors, coupled with the equal opportunities for education and the competitive system for entering science, have brought about a lasting tendency of (quantitatively) equal participation of women in the various fields of scientific knowledge. At the stage of entry into science, the positive cumulative effect of the entire combination of factors influences the principle of equality.

Quite different is the picture of advancement in science. Here other kinds of laws are in operation. In this 'other world' even general social laws seem to function in 'an upside down' way. The fact that accelerated industrialisation has left behind the sphere of services directly affects the time budget of women scholars. The quality achieved in economic activity and education has reversed the concepts of 'the fair sex' and 'the strong sex', so that now women bear the double burden, ironically explained as 'attained emancipation'.

In order to be able to preserve their identities and to be recognised as personalities, women are not likely to betray their nature. Since Bulgarian women scientists will not concede to being deprived of the joy of having children, they would rather get rid of the chains of marriage. The illusions associated with the 'almighty collectivism' are disappearing. Scientific career promotion of significance, the organisation of scientific research activities, the atmosphere and care for the young people within the scientific organisations, all play important roles. However, in the final analysis, everyone remains alone on their own road to success in science.

Women in science are like the lonely long-distance runners: not only is their track blocked at all bends with barriers that are met by all scholars, but also with some barriers specific only to women scholars. This does not imply that society is incapable of accelerating this movement. However, so far its measures have been of a rather general and indirect nature. The assistance to women within the scientific organisations themselves is bound to be more effective: the behaviour of women who have already attained positions of responsibility and of men who have abandoned the historical burden of scepticism, are particularly relevant.

In conclusion, it is possible to think of the advancement of women in science as resembling 'the loneliness of the long-distance runner'. It requires talent, endurance and perseverance. But in fact 'there is no easy way in science and an access to its bright peaks can be had only by those climbing up without fear of its precipitous paths'. (Marx, vol. 33:368) In this context, are not the numerous hindrances to women in science the necessary trial and the guarantee for a real breakthrough in science?

References

ANANIEVA, N., 'Women's Access to Science: Factors and Obstacles (Bulgaria)', paper presented at the Unesco Expert Meeting on Factors Influencing Women's Access to Decision-Making Roles in Political, Economic and Scientific Life, Dubrovnik, December 1984.

Balgarskata Akademia na naykite sled Deveti septemvri 1944 (Bulgarian Academy of Sciences after 9 Sept. 1944.) Izdatelstvo na BAN, Sofia, 1958.

BLAGOEV, D. AND G. DIMITROV, *Za jenata i semeistvoto*. (On Women and the Family.) Sofia: Partizdat, 1979.

DIMITROV, KR., *Balgarskata Inteligenzia pri capitalisma*. (Bulgarian Intelligentsia under Capitalism.) Sofia: Nauka i Izkustvo, 1974.

DINKOVA, M., *Sozialen portret na balgarskata jena*. (A Social Portrait of the Bulgarian Woman.) Sofia: Profizdat, 1980.

HOLOV, P., *Balgarskite Istorizi*, Bio-Bibliografski narachnik (Bulgarian Historians, Bio-Bibliographies Manual.) Sofia: Nauka i Izkustvo, 1981.

Istoria na Sofijskia Universitet, Kliment Ohridski (History of the 'Kliment Ohridsky' Sofia University). Sofia: SU Publishing House, 1988.

MARX, K. AND F. ENGELS, *Ludwig Feuerbach i krajat na klassicheskata nemska filosophia, tom 21*, (Ludwig Feuerbach and the End of Classical German Philosophy.) Sofia: Partizdat, 1967.

MARX, K., *Pismo do Morris Lachatre*. (A Letter to Morris Lachatre.) London, 18 March 1872, vol. 33, Sofia, 1967.

Medizinska Akademia-Plovdiv (Medical Academy – Plovdiv), 1945–70. Izdatelstvo Hristo G. Danov, Plovdiv, 1971.

PUNDEVA, AL.-VOINIKOVA, *Balgarkata prez Wazrajdaneto*. 'Izdatelstvo na Sauza na balgarskite jeni' (The Bulgarian Women in the Renaissance.) Sofia, 1940.

Statisticheski godishnik 1978 (Statistical Yearbook 1978). Sofia: ZUJ, 1979.

Statisticheski godishnik 1987 (Statistical Yearbook 1987). Sofia: ZUJ, 1988.

Sto godini Balgarska Akademia na naukite (1869–1969), Hundredth Anniversary of the Bulgarian Academy of Sciences, vols I, II, III. Sofia: Izdatelstvo na BAN, 1972.

Unesco, *Participation of Women in Research and Development*. A statistical Study. Division of Statistics on Science and Technology. Office of Statistics, Sept. 1980.

6

Soviet Women in Science

Vitalina Koval

Introduction

Women's involvement in intellectual activities everywhere in the USSR, the rapid growth of their professional and educational level (women make up sixty-one per cent of persons with higher and specialised secondary education), and new opportunities to apply their efforts and knowledge in technology have resulted in the desire of a large number of women to do scientific research. Moreover, the relatively slow rate of scientific-technological progress and the limited use of technological innovations in industry narrow the sphere where people with higher education, especially women, can apply their knowledge and skill. In science, however, they can reveal and broadly use their knowledge and potentialities. Thus, it is no wonder that during the 1970s and 1980s there has been a significant rise in the number of women scientists, who comprise forty per cent of all scientists at present.

The democratisation of society in the *perestroika* period is accompanied by the large-scale involvement of all women and above all, women doing intellectual work in political activity. The National Conference of Women held in Moscow in January 1987 discussed ways of recreating women's councils. Over a brief period, 160,000 such councils with a membership of 1,400,000 have been set up in the country. Women's councils have been established in all the Academy subdivisions to help women scientists solve their problems. Women scientists took an active part in the election campaign to nominate candidates to the Congress of People's Deputies from the USSR Academy of Sciences. They vigorously supported nominees with progressive views, and one woman academician, T. Zaslavskaya, was elected a deputy from the Academy of Sciences (from twenty-four deputies). In the new Parliament, women make up 17 per cent of the members, among whom there are many scientists. A solution to urgent women's problems and the role of women in society depends on how actively women members of parliament uphold their interests. Recognising the acuteness of women's problems, their social and political significance, the new Supreme Soviet formed a committee on the position of women, the protection of mothers, children and the family. Work is under way to draft a special comprehensive programme, 'The Working Women's Labour, Rest, and Life'. This programme is to plan measures to create working, rest, and living conditions that would enable women to successfully combine their two roles, that of the worker and that of the mother. Women should be given more benefits as compared with men to ensure women's equality, not only formally but also taking into account the specific features of their problems. Women scientists have been taking the most active and direct part in drafting this programme since all these points need a profound scientific analysis and study.

The government has also set up a commission on women's problems. The task of this body is to prepare guidelines for solving these problems, determine the necessary expenditures, appoint those responsible for these measures, and set deadlines. Women scientists will be actively involved in the work of the commission. The success of *perestroika* in society as a whole depends on how actively the potential of women, especially scientists, will be used.

Background Information

Over the seventy years which have passed since the 1917 Socialist Revolution there have been fundamental changes in the educational, cultural, and technical standards of the population in this country. Women were made fully equal with men in all fields right after the Socialist Revolution. When laws on the protection of mothers and children were enacted, a ramified network of schools and pre-school establishments was set up, and illiteracy was eradicated. A compulsory education system was introduced.

Women were given all possibilities to acquire any profession at all levels, including access to institutions of higher education and vocational training schools, on an equal footing with men first on the primary and then on the secondary educational level. The large-scale enrolment of young women at educational institutions enabled them to learn different professions in all fields including the technical ones. At vocational training schools, women are able to choose between nearly 1,000 occupations from about the 1,500 offered, with the exception of occupations dangerous for women's health.

The high level of women's employment (92 per cent of women of working age are employed) is the key to studying the position and role of Soviet women in society. There have been remarkable achievements in the professional training of women and the use of women in public production. Women make up 51 per cent of the work force. They represent 58 per cent of engineers, 67 per cent of physicians, 87 per cent of economists, 89 per cent of book-keepers and 61 per cent of specialists with higher and specialised secondary education. Their number has risen 24-fold compared with 1940.

Scientific and technological progress, large-scale automation, computerisation and robotics have called for a large number of skilled personnel and specialists in new professions: programmers, coders, perforators and basic and auxiliary equipment operators. Women comprise from 50 to 60 per cent of these workers.

Among the students at higher education establishments young women comprise a little more than 50 per cent. However, at the technical colleges, women make up about 40 per cent of specialists training for industry, construction and transport.

Several generations of women have grown up under the Soviet government, and their orientation in life aims at work in public production. They are guided not only by economic incentives but also by the desire for self-expression and self-accomplishment. Work gives women moral satisfaction and economic independence from men. However, in coping with the problem of involving women in public production successfully, society is confronted by a number of serious problems of how women carry out their main childbearing function and family duties.

Although the protection of the interests of mothers and children is part of our state policy, nevertheless women and families have been socially unprotected. One was not accustomed to speak about this, but *perestroika* and *glasnost* have exposed the serious problems that are now extensively discussed by the mass media as well as by government and party officials. For the first time in many years, women's problems are seen as acute social and political issues, requiring a fundamental approach and a solution.

Unfortunately the available women's corps of experts has so far been used inadequately; professional careers have been problematic without the guiding support of society. The creation of conditions ensuring women's active participation in scientific, technological and industrial activities should be viewed in our time as a necessary condition for social progress as a whole.

What Does it Take to Become a Scientist?

There is a special system for becoming a scientist in the Soviet Union. Any man or woman with higher education qualifications can get employment at scientific research centres as a laboratory assistant – the first level of scientific jobs.

The second stage is full or correspondence post-graduate study[1] (three years off the job or four years on the job). During this time, a post-graduate student gets comprehensive theoretical training in his or her speciality, writes a dissertation and defends it before a panel of scientists. The first post-graduate course was opened at the USSR Academy of Sciences in 1929, and 25 per cent of the first post-graduate students were women.

Upon graduation and the successful defence of his or her dissertation, a post-graduate student is granted a lower level doctorate(Ph.D.). Good progress in science, successful research experience of ten to twelve years, provide an opportunity to apply for a two–year course to prepare a higher doctoral dissertation. After the successful defence of the dissertation, the

---

1. Admitted to post-graduate study are persons with complete higher education under thirty-five years of age, who have revealed ability or inclination for scientific research through an examination. Post-graduate students, studying by correspondence, are entitled to an additional fully-paid one-month's leave.

Doctorate degree is conferred. It usually takes eighteen to nineteen years to obtain a Doctorate. The availability of a large number of publications before the defence of the dissertation is envisaged. The Ph.D. and the Doctorate are both conferred through voting in a secret ballot by noted scientists in a given field. A scientist with a Doctorate degree can aspire to honorary Academy membership (Corresponding Members and Academicians).

Do Women Really Have Equal Opportunity in Science?

Formally, women enjoy the same rights and opportunities for scientific activity as men. The number of women scientists rose from 128,700 in 1960 to 598,100 in 1986 or nearly five-fold. But in reality women's position in science continues to be unequal with men. Comparing the roles of men and women in scientific research shows that there is a hierarchical difference in the division of scientific labour between the sexes. In spite of the fact that women have improved their position in the scientific hierarchy during the last three decades, their situation continues to be unequal because many obstacles exist in scientific institutions. The fact is that men occupy the decision-making positions and determine what kind of work should be done by women. By and large women hold second-rate positions in science, their choice of independent research topics is limited, and they carry out more auxiliary scientific work.

The position of women in scientific institutions is largely determined by their participation in scientific studies. They are more often authors of articles, development projects, and separate chapters in collective monographs and less frequently independent authors of books. At one institute of humanities under review, only 62 (23 per cent) of 269 monographs were written by women. The lion's share of scientific analytical work is done by men.

Scientific work is distinguished by many specific features hampering women's involvement in this sphere of public production, especially their promotion opportunities. It is here that the contradiction between the two roles that women play – work and family duties – is most acute. Maternity and family obligations, which are shouldered mainly by women, often poor housing conditions, inadequate services, a shortage of consumer goods and lack of desire on the part of men to share housework with women create obstacles to women's careers.

The contradictions between women's professional activity and family duties and maternity have become increasingly acute. Women have to work twice as hard physically and mentally and sacrifice their leisure and intellectual growth. Additionally there are some subjective obstacles, such as prejudices which still portray women as second-rate personnel in science, and different obstacles on the part of administration during elections or appointments to leading positions. In order to achieve the same career as men, women must work much harder, be cleverer, more capable and more educated.

Initial conditions for scientific work upon graduating from higher education institutions are equal for young men and women. However, for women the period of scientific maturity, as a rule, coincides with marriage and motherhood. During the eighteen months of maternity leave, science moves forward and time is needed to catch up with what has been missed.[2] The gap between men and women in defending the dissertation thus increases. Among women post-graduates, only 26 per cent defend their dissertation in time (after three years of post-graduate course); the figure for men is 40 per cent. The length of time between the defence of Ph.D. and the degree of Professor is five years longer for women than for men (16.5 years and 11.5 years respectively).

Married women often prefer a post-graduate course by correspondence, continuing to work without a reduction in pay. This increases the gap still more. Studying at a post-graduate course by correspondence entitles a person to one month of additional paid leave. Women with family and children, as a rule, are unable to carry out research in time after work (after 4 years of post-graduate course of study). Men spend twice as much of their time after work on research as women, who spend twice as much of their time after work on household duties (38–40 hours

---

2. They are entitled to 112 working days of fully-paid leave plus six partially-paid months, as well as unpaid leave until the child is one and a half years old. The defence of a dissertation is postponed for this term and even longer, if research is connected with laboratory experiments.

a week). Despite all the obstacles facing women, the number of women who defended their dissertations over the last 25 years soared from 28,800 in 1960 to 111,100 in 1980 and to 132,000 in 1986. This nearly 9-fold increase points to a new growth in women's educational standards in terms of quality. In 1986 women comprised 28 per cent of all Ph.D.s (483,000). However, the higher the scientific degree and title, the fewer the women holders. Among all professors, only 13 per cent are women. Among academicians, corresponding members, and full professors, there are 11.3 per cent of women across the Soviet Union, but among assistant professors their number reached 24 per cent in 1986 as compared with 17 per cent in 1960. The trend in the development of the position of women on different levels of the structure of the scientific community is shown in Table 6.1.

Being juniors in job, rank, and degree, women are less rewarded for their work than men. The salaries of laboratory assistants and junior scientific researchers are 120–150 rubles, which is 7–8 times less than the salaries of Academicians and Corresponding Members. Therefore, it is not surprising that women often voice dissatisfaction with their work and position and seek to upgrade their working status, to get a higher scientific degree and title. Most successful in science are single women, who have decided to devote their life to science. Only a few married women with children defend their dissertation for Professor or Full Professor degrees and become senior scientific researchers. The majority remain junior scientific researchers and often laboratory assistants for the rest of their life.

The Structure of Women Scientists' Employment: A Case Study

In the Soviet Union the majority of scientists are concentrated in the Academy of Sciences. The Academy consists of more than 150 institutes in different fields, with about 300 subdivisions, employing more than 60,000 researchers. Each Republic has its own Academy, and the USSR Academy of Sciences has its headquarters in Moscow. Table 6.2 shows the distribution of the total number of scientists employed by the USSR Academy of Sciences, and the proportion of women according to rank. In

Vitalina Koval

Table 6.1: The proportion of women among scientists (in thousands) in different positions, 1960–86

	1960 N	1960 %F	1970 N	1970 %F	1980 N	1980 %F	1986 N	1986 %F
Scientists (including universities and research institutes)	354.2	36.3	927.7	38.8	1 373.3	39.9	1 500.5	39.8
of these scientists with:								
– Doctorate degree	10.9	10.1	23.6	13.1	37.7	14.1	45.7	13.3
– PhD (candidates)	98.3	29.2	224.5	27.0	396.2	28.0	472.8	28.0
Those with scientific titles:								
– Academicians, corresponding members, full professors	9.9	0.7	18.1	9.9	27.4	10.9	31.8	11.3
– Assistant professors	36.6	17.1	68.6	21.0	110.7	23.7	134.1	25.0
– Senior researchers	20.3	28.6	39.0	25.1	66.0	22.6	78.6	22.0
– Junior researchers	26.7	50.9	48.8	49.8	41.0	46.6	38.4	44.8

Source: National Economy of the USSR in 1987. *Statistical Yearbook*, Moscow: Finances and Statistics, 1988, p. 27. (In Russian).
N = Total number.

1988 women comprised 37.5 per cent of the more than 60,000 scientists of the Academy.

As Table 6.2 shows, in the Academy of Sciences women are not on a par with men. The higher the title and office, the fewer the women. Although attitudes to women in science have changed over the last seventy years, women still hold few managerial posts. In order to get a better picture of the career obstacles and opportunities of women scientists and their position in different branches of science, seven institutes of the Academy of Science were chosen for a case study. Three of the institutes deal with the technical sciences, two with the natural sciences, one with the humanities and one with the social sciences. These institutes were chosen because they have a large

Table 6.2: The structure of the scientific labour force in the USSR Academy of Sciences in 1988

	Total number	% women
Academicians and Corresponding Members	550	1.6
Holders of a Doctorate degree	6 281	14.9
Holders of a Ph.D. degree	27 817	34.4
Full Professors	1 865	7.7
Assistant Professors	758	15.0
Senior Scientific Researchers	8 704	24.5
Junior Scientific Researchers	8 008	47.8
Scientists without any degrees	26 702	41.7

Source: Information from the USSR Academy of Sciences, 1989.

percentage of women among scientists.

Data from the seven institutes show that most women are on the scientific staff of the two institutes of natural scientists (61.5 and 55.2 per cent). Next, most women are at the institutes of humanities and social sciences (57 and 44 per cent respectively). The three institutes of technical sciences are significantly different; here women scientists comprise 20, 22, and 36 per cent. Thus, women seem to be more inclined towards the humanities and natural sciences than the technical sciences.

One common trend for all is the small number of women among professors. In the institutes of humanities and social sciences as well as natural sciences, their proportion ranges from 11 to 29 per cent. In technical institutes, their share among professors is approximately one-third less (from 3 to 8.6 per cent).

There is a similar trend among women with Ph.D.s, although women have achieved much as far as the technical field is concerned. In natural sciences, women with Ph.D.s comprise an average of 55 to 58 per cent – the highest level in the scientific fields under review. In the institutes of humanities and social sciences, the share of women with Ph.D.s is 33 and 51 per cent respectively. In technical institutes, the proportion of women having Ph.D.s is less and ranges from 10 to 23 per cent.

There is an interesting trend among scientific workers without a scientific degree. The highest proportion of women without a degree is in the institutes of humanities and social sciences (71 and 78.8 per cent) and in the institutes of natural sciences (66.9 and 75.3 per cent). In the technical sciences, the percentage of women without a degree is small (22.7 and 26.9 per cent), an average of four times less than for men. This is explained by the fact that women make up only one-quarter or one-fifth of all scientific staff in these institutes. There is every indication that preference in employment is given to persons with scientific degrees (above all this concerns women).

Regrettably, there is not a single woman among the thirty academicians or corresponding members of the seven institutes. Among 153 full professors only eight are women.

Thus, although the number of women scientists is relatively high and has gone up considerably as compared with the previous period, women scientists continue to perform mainly helpful functions of laboratory assistants and junior scientific researchers, and their participation in positions of leadership is extremely low. This is a reflection of the general picture in the country as a whole: while comprising 61 per cent of the population with higher and specialised secondary education, women hold only 7 per cent of leading posts.

Trends in Science Connected with Perestroika

Now, when society is being renewed through democratisation and openness, significant changes are taking place in science and its organisational structures. A new system of elections to leading posts has been introduced at all levels in the scientific institutions of the Academy of Sciences. During 1986–9, new institute directors and scientific department chiefs were elected. The elections were carried out on a democratic basis with all the institute employees taking part in the voting by secret ballot. All the organisations, scientific departments or individuals were able to nominate their candidates to leading posts.

The election returns showed a relative rise in the number of women holding key positions. Six women (or 2 per cent) were elected as institute directors to some of the 293 scientific subdivisions of the Academy. As a positive example, the election of a woman to the post of director of the newly-created institute on the problems of demography can be cited. Of 504 candidates, twelve women (or 2.2 per cent) were elected as assistant directors. There are more women among department, laboratory, and group chiefs. The average figure for the Academy of Sciences is 11.6 per cent. In the technical institutes under review, women account for 3 per cent among the chiefs of the laboratories; and in natural science institutes 10 per cent. At the two institutes in the humanities and the social sciences, the proportion of women among elected group chiefs is 24.1 per cent and 8 per cent respectively. Among the scientific secretaries at all academic institutions, women make up 32 per cent, which testifies to the recognition of women's organisational abilities.

What is surprising is that at most institutes women were not elected to the steering bodies even where women make up the majority of the electorate. How can this be explained? One explanation may be that there are outdated stereotypes about the role and place of women in society and science, stereotypes that even women are unable to overcome in their consciousness. Quite often women vote against women candidates nominated for major offices, since they regard them as being more capricious and harder to deal with. The results of the steering body elections for academic scientific subdivisions show that little has changed, despite the decisions adopted by the party and government to nominate women to leading posts. This process has been slow, being constantly delayed while the gap between the number of men and women in leading posts remains almost unchanged. The share of women in the overall number of scientific institution chiefs in the system of the USSR Academy of Sciences is miserable (2.2 per cent).

Elections to scientific councils in the seven institutes show the same trends as the election to steering bodies. There is a slight increase in the number of women. Among members represented in scientific councils of technical institutes women make up 3 per cent, in natural science institutes 9.4 per cent, in humanities and social science institutes 9.5 per cent. Thus, women are most successful in natural sciences and the humanities.

An Example from the Social Sciences

The institute specialising in the social sciences was taken for a more detailed examination of women's position. Women there comprise 57 per cent of the scientific staff. They are specialists in various fields – economists, philosophers, sociologists, students of politics, ecologists and historians studying the laws of the development of society from various angles.

Among researchers with a scientific degree 51 per cent are women; of the 33 doctors of science 21 per cent are women. Women comprise 78.7 per cent of researchers without a scientific degree.

Four women were elected to the new scientific council which is composed of twenty-seven members. All of them are doctors and two of them are professors. The representation of women in the council has risen from 8.8 per cent to 15 per cent. On the other hand, there are no women among directorate members, or among department or section chiefs. All seven of the group chiefs are women.

For more detailed analysis a survey was made among the seven leading women scientists of the institute (professors and doctors of science) and of twelve other women with Ph.D. degrees. All of the seven leading scientists defended their thesis for a doctorate at or over the age of fifty, with the exception of one woman who defended her thesis at the age of forty-three (in comparison with forty to forty-five years for men in general). The time gap between the Ph.D. degree and the Doctorate ranged from ten to thirty years (as compared with ten to fifteen years for men).

One of the seven women is single, has never been married, and has devoted all her life to science. Two of the women are divorced, and four of the women are married with children.

The seven respondents mentioned the following factors that promoted their careers:

1. Sound training for scientific research at university.

2. Purposefulness, interest in scientific research.

3. Teaching experience.

4. Creative atmosphere in the scientific team where they worked.

5. Transfer to creative scientific work within the Academy of Sciences.

6. Assistance by parents to look after a child.

7. Assistance by husband and parents in housework.
8. The husbands also being scientists (all four of the married women's spouses are scientists).
9. Encouragement in the scientific team and from the management, material support from the Academy of Sciences during post-graduate study.

The following were mentioned as negative factors obstructing women's scientific careers:

1. Poor living conditions, especially at the early stages of the formation of a woman's scientific career.
2. Difficult family relations (conflicts).
3. Much time devoted to household duties.
4. Among those who worked in the outlying regions – remoteness from the central archives, the lack of scientific literature and periodicals, especially in foreign languages, the shortage of contacts with leading specialists.
5. Sceptical, prejudiced attitude towards women, obsolete stereotypes in the determination of the role and place of women in society and science, neglect of competence and professionalism of women scientists in scientific spheres.
6. The incomplete use of women scientists' potentialities and their inadequate promotion.
7. Difficulties with publications due to the absence of appropriate printing facilities.
8. In pre-*perestroika* times, scientific work was hindered by the system of special archives, closed archives, strict limitations on personal contacts with foreign scientists, the restriction of international ties and participation in international events (symposia, conferences).
9. Poor technical equipment, the shortage of xerox machines, printers, computers, etc.

These conclusions were also endorsed by the other twelve women scientists interviewed (all under forty years of age).

These women are most of all worried about the problems of bringing up children, leisure time, and the shortage of basic necessities. All of them have children (not more than two). The greatest difficulty for them is to combine two roles: that of the mother and that of the scientist. Because of household problems, they spend about forty hours a week on household work. Their leisure time amounts to only two hours twenty-four minutes a day. Of this, they spend on the average one hour thirty-nine minutes on cultural activities (television, radio, going to the cinema or theatre, reading a newspaper), eleven minutes a day to entertain guests, eight minutes a day to visit restaurants, cafés, bars. Only sixteen minutes a day are spent on child care.

Changes are taking place in the public consciousness, although slowly. The only model determining women's equality and freedom in Soviet society that has guided us after the Socialist revolution, has been a model worked out in line with Marxism. This model has not solved the women's issue in full. In recent years, criticisms have been voiced over the position of Soviet women, and alternative views on the role of women in society have been published in the press. One thing is clear: the complete involvement of women in public production without creating favourable living conditions to combine the two roles did not produce the desired result Marxist philosophers dreamed of. Scientists are facing a great and fundamental task to work out a scientific and theoretical concept of solving the problems facing Soviet working women in the period of democratisation. So far such work has been sluggish, although the Presidium of the Academy of Sciences has instructed many institutes of social sciences to include 'women's problems' in their research plans.[3]

The surveys of women's problems in all fields demand deep and serious theoretical and social studies, especially under the new conditions of democratisation and *glasnost*. Unfortunately research in this field is only at the very beginning.

---

3. The book *Women in the Contemporary World* (published 1989) was prepared jointly by thirty-eight scientific researchers at various institutions of the Academy of Sciences and other research centres.

References

Narodnoe khozyaistvo SSSR v 1987 g. Statisticheskiĭ ezhegodnik. (National economy of the USSR in 1987. Statistical Yearbook). Moscow: Finances and Statistics, 1988.

Narodnoe obrazovanie, nauka i kultura v SSSR. (Education, science and culture in the USSR). Moscow, 1971.

Naselenie SSSR za 70 let. (The population of Russia over 70 years). Moscow: Nauka, 1988.

RASHIN, A., *Naselenie Rossii za 100 let.* (The population of Russia over 100 years). Moscow, 1956.

SERGEEVA, G.P., *Professionalnaya zanyatost zhenshchin: problemy i perspektivy.* (Women's professional employment: problems and prospects). Moscow: Economy, 1987.

Sovetskaya zhenshchina: grud, materinstvo, semya. (The Soviet Woman: Work, maternity, family). Moscow: Profizdat, 1987.

Women in the Contemporary World. Moscow: Nauka, 1989.

Zhenshchiny v SSSR. 1988. Statisticheskie materialy. Goskomstat SSSR. (Women in the USSR. 1988. Statistics). Moscow: Finances and Statistics, 1988.

7

Women, Science and Politics in Greece: Three is a Crowd

Ann R. Cacoullos

Introduction

The Greek case, which has not been systematically investigated so far, certainly exhibits some of the abiding trends which have been observed and studied in other societies. Thus, in the university faculties, women are heavily concentrated in the fields of the humanities and social sciences. In independent careers, where no research is involved, there is a high percentage of women in pharmacology, dentistry, law, and most recently, architecture. These occupations are chosen by Greek women because they permit maximum flexibility within the 'double-burden' scenario and at the same time provide high satisfactory remuneration. On the other hand, too few Greek women are found in the fields of the natural and medical sciences, engineering and technology; moreover, as we shall see below, women in the university hierarchy are concentrated in the lowest academic posts.

The recent substantial increase in the number of young Greek women entering the universities and enrolling in traditional 'male' fields, for example mathematics and the natural sciences, does not so far mean that greater opportunities exist for the successful pursuit of careers in these areas. In addition to the familiar constraints widely documented in the literature (e.g., familial, attitudinal, etc.), Greek women confront certain society-specific obstacles which contribute substantially to their on-going marginalisation in science. Our aim is to illustrate some of these in this study.[1]

---

1. I acknowledge with great thanks the help of T. Cacoullos of the Greek

Ann R. Cacoullos

Greek Women in Higher Education: Students and Faculty

There has been a remarkable change in balance between the two sexes in the Greek university student population in recent years. In the 1950s, out of four students, three were men, one a woman; in the late 1970s, out of five students, three were men, two were women. (Lambiri-Dimaki, 1986:103) Today, the late 1980s, in the largest urban area of Greece, metropolitan Athens, out of five students, three are women, and in Salonika, Greece's second largest city, more than half of the student university population (54 per cent) is composed of women.[2] These figures, however, should not precipitate a celebration of women's place in scientific fields. While the female university population increases both in absolute numbers and percentage-wise (compared to male students) – of the 28,157 students admitted to universities in 1987/8, 55 per cent or 15,504 were women – the highest percentage of women students (85 per cent) is enrolled in the faculties of literature, philology and languages, followed by a smaller number (58 per cent) in law and economics, education and physical education. The lowest percentage (27 per cent) is observed in engineering departments. Women comprise about 10 per cent in electrical and mechanical engineering, however, over 50 per cent in architecture. In the fields of the natural sciences and mathematics, women compose 40 per cent of the student population in mathematics, over 50 per cent in biology and chemistry and close to one-third in physics. In the medical or health sciences, over 50 per cent of the student population is female (see Table 7.1).[3]

There is no doubt that Greek women are converging in hordes on the university; what is doubtful is the future cash and career values of their university education. The fall in male enrollment can be attributed to an increasing realisation that these values

---

Statistical Institute and A. Chatziparadisis of the General Secretariat for Research and Technology (GSRT) in the collection of most recent data regarding women in science.

2. Figures on women students have been compiled for this study from unpublished data of the National Statistical Service of Greece.

3. While the number of women students entering the medical sciences is increasing compared to the number of men, women continue to concentrate in the fields of dentistry, nursing and pharmacy since, as already stated, there are greater career opportunities for women in these areas.

Table 7.1: Distribution of female students admitted in 1987/8 by field or department in three main universities

Field	Athens	Salonika	Patras	Total	%
Theology	286	230	–	516	4.4
Law	1 208	802	–	2 010	17.2
Medical Sciences	347	282	80	709	6.1
Humanities	1 826	1 233	–	3 059	26.2
Natural Sciences Mathematics	439	352	240	1 031	8.8
Pedagogy	–	–	–	–	–
Physical Education	897	270	282	1 449	12.4
Engineering	256	253	136	645	5.8
Agriculture		205	–	205	1.8
Business	445	–	–	445	3.8
Social Sciences	682	–	–	682	5.8
Management	408	361	–	769	6.6
Total	6 844	4 061	738	11 643	100.0

Source: Compiled from unpublished data of the National Statistical Service of Greece.

Remarks: figures for male students admitted in this period are not yet available.

are progressively falling; the rise in women's enrollment can be attributed to the still pervasive trend towards university education, come what may, and the contemporary tight job market. Greek women completing the Lyceum, or the final three years of high school which are mainly college preparatory, have very few opportunities for jobs in Greece; one way of postponing the vacuum is to enroll in the university and enter the long queue of university graduates waiting to be appointed to secondary school teaching or join the civil service.

While we have no formal figures to date, we do know that very few Greek women will be able to undertake advanced education abroad which will qualify them for scientific careers in the university and research institutes. Apart from this abiding fact, there is another discriminatory phenomenon, most recently visible and illustrated by the distribution of the student population in the newly developed academic programmes of computer science and information technologies at the universities of

Patras and Crete and the School of Business and Marketing Sciences in Athens (ASOEE).

On the basis of data we have collected, we note a significant increase in the number of women admitted to the ASOEE programme of statistics and informatics, and a curious wavering in women's percentages in the Patras programme. (Cacoullos, 1988:81, Table 2.8) In both Patras and Crete, the course of study to which students are admitted is a clearly demarcated department of information sciences, and overall we ascertain a constant average of 23 per cent of women students in the four-year period, 1984–8. In ASOEE, on the other hand, the course of study is part of a mixed bag of business statistics and computer science, and we note a significant increase of women students, from 37 per cent in 1983–4 to 61 per cent in 1987–8 which surpasses the percentage of male students for the same period.

The example provided here is instructive: women students choose, in the main, to study for low-level technology training programmes (it is an acknowledged fact that entrance into the ASOEE programme is easier), whereas male students choose, in the main, to sit for the more demanding examinations of the first category where the probability of admission is lower, for example at Patras and Crete. One could argue that women are responding to societal expectations, always lower for them, but it can also be urged that women students are being realistic about the opportunities open to them. The scientific job market in Greece is limited, and women continue to be second to men. This is clearly a structural problem which reinforces an insidious sort of discrimination against women: they are encouraged to study and to plan careers, but 'too much' study and career planning are undesirable from the point of view of the needs of the Greek social formation. Thus, while no longer considered unintelligent or less competent than men, Greek women students are still being absorbed into the less demanding scientific programmes even as their numbers increase in the university student population as a whole.

Moreover, as has been noted, if men are generally lagging behind women in Greek university student populations at the moment, it is not because they are less 'serious' or losing interest but because many are reconsidering the scientific priorities in the country, a luxury not yet available to women who are expected to be thrilled about university entrance, which is another kind of obstacle to career planning.[4]

In one of the very few studies of Greek women in university faculties, the ongoing imbalances which characterise the careers of women in academe are confirmed. In the 1970s, the percentage of women among full professors throughout Greece was below five (4.8 per cent, 1972–3, 4.9 per cent, 1979–80); in the remaining teaching staff, or lowest positions, in 1972–3 women constituted 63.7 per cent and in 1979–80, 59.8 per cent. (Eliou, 1988:10, Table II) That women constituted 35.7 per cent of the total teaching staff in 1979–80, as also reported in the Unesco *Statistical Yearbook*, 1982, should not impress, given the distribution of women faculty in the teaching hierarchy.

The present situation, while showing certain advances, has not substantially altered the balance. Imbalances have become almost universal constants and all too familiar. Women are under-represented in the occupation of university professor, they are under-represented in those academic fields having the highest prestige which also provide greater opportunities to combine teaching with independent research outside the university. They are not occupying influential administrative posts. Finally, they show such an unequal distribution on the various faculty levels that it is clear that their careers and promotion on the hierarchy have been obstructed, interrupted, or evolve at a much slower rate than that of their male colleagues.

Recently it has become clear that the number of university faculty women has not increased much beyond the initial advance achieved in 1975.[5] More importantly for our purposes here, their numbers continue to vary inversely with the academic prestige attached to the faculty position (rank) and academic field. Thus, the highest percentage (15 per cent) of women professors is in humanities faculties. With respect to all disciplines and all faculty ranks, there is one woman among five faculty members: one woman among six in engineering and technology faculties and one among three in humanities (distribution in the two major universities, Athens and Salonika).

---

4. Recent attempts to introduce career planning programmes in secondary schools have not been successful. Thus the main aim for most young Greek persons is university entry which virtually delays or suspends career planning.
5. Figures on university faculty women in the late 1980s have been compiled for this study on the basis of unpublished data of the Greek Ministry of Education.

The latest available data which include fields in the health sciences indicate that over five per cent of university professors in the sciences are women: in biology, 12 per cent are women while there is only one woman professor (among 180) in the departments of engineering and technology in all polytechnic schools. Summarising, just over 20 per cent of all science faculty members (2,709) in all ranks are women, with about 25 per cent in the medical sciences. If the health sciences are excluded from the general picture, then just over three per cent of university professors in the sciences (457) are women. There are no women among the dozen or so professors in the four recently instituted departments of computer science (see Table 7.2).

We have focused on the rank of professor in the main because it is still true in Greece that a successful career in the sciences is associated with an appointment in the university, and specifically to the rank of professor. Today this is changing, with attempts to reformulate science and technology policy in Greece, and ranking in research institutes. For the moment, however, it is the case that an appointment to the highest rank of the university system is highly desirable, though less than it used to be ten years ago; it is a career that provides job and financial security, with retirement benefits that are among the best in world-wide social security programmes. Election to the ranks of professor and associate professor presupposes a doctoral degree as well as publications. Greek women have not fared well in this career area for the widely acknowledged reason that the majority of young women in Greece are inhibited, if not outrightly discouraged, by their families from studying abroad, which they would have to do in order to satisfy the qualifications for a university appointment. It has been remarked that a peculiar characteristic of the Greek university is that it does not reproduce itself, that is, the overwhelming majority of its faculty has been educated abroad. (Frangoudaki, 1985:200) While not specified by law, in actual practice those who have obtained a doctorate abroad are favoured over those who have a Greek doctoral degree, especially in scientific fields. The consequences of this entrenched practice are obviously mainly negative for women.[6]

6. The present ranking system: professor, associate professor, assistant professor and lecturer, recently instituted by the new University Constitution Law

Table 7.2: Science faculty by department or university school, rank and sex, 1985/6

Department or School	Athens Profs T W	Athens Others T W	Salonica Profs T W	Salonica Others T W	Ioannina Profs T W	Ioannina Others T W	Crete Profs T W	Crete Others T W	Patras Profs T W	Patras Others T W	Thrace Profs T W	Thrace Others T W	All Greece Profs T W	All Greece Others T W
Natural Sciences														
Biology	7 1	47 20	10 –	50 20	~ ~	~ ~	3 1	5 2	7 1	33 11	~ ~	~ ~	27 3	135 53
Geology	6 0	46 11	4 –	37 6	~ ~	~ ~	~ ~	~ ~	4 –	11 2	~ ~	~ ~	14 0	94 19
Physics	11 1	89 24	19 1	66 3	8 0	32 5	11 0	17 0	11 –	28 5	~ ~	~ ~	60 2	232 37
Mathematics	8 –	30 10	10 –	31 13	9 0	28 2	7 1	14 2	12 –	24 5	~ ~	~ ~	46 1	127 32
Total	32 2	212 65	43 1	184 42	17 0	60 7	21 2	36 4	34 1	96 23	~ ~	~ ~	147 6	588 141
Technical and Engineering	69 1	257 58	53 –	156 38	~ ~	~ ~	1 0	2 0	33 0	54 1	24 0	35 4	180 1	504 101
Medicine	31 –	492 137	37 5	373 79	19 0	48 11	~ ~	~ ~	19 0	45 13	2 0	15 0	108 5	973 240
Dentistry	10 2	93 36	5 –	57 16	~ ~	~ ~	~ ~	~ ~	~ ~	~ ~	~ ~	~ ~	15 2	150 52
Pharmacy	3 –	21 12	2 –	15 3	~ ~	~ ~	~ ~	~ ~	2 0	1 0	~ ~	~ ~	7 0	37 15
Total	44 2	606 185	44 5	445 98	19 0	48 11	~ ~	~ ~	21 0	46 13	2 0	15 0	130 8	1 160 307
Medical Sciences														
Total	145 5	1 075 308	140 6	785 178	36 0	108 18	22 2	38 4	88 1	196 37	26 0	50 4	457 15	2 252 549

Source: Ministry of Education

Other = Associate, Assistant, and Lecturer (Before 1985 only Assistant and Docent)
~ indicates non-existent department

The major reform of higher education initiated in 1982 with the passage of a new University Constitution Law – *Nomos-Plaicio* – has had no substantial impact on the situation of women seeking careers in university teaching, nor, indeed, was it intended to address issues of gender imbalance in university hierarchies and teaching ranks. The new law eliminated the position of the hitherto all-powerful Chair – *Edra* – to which all others (eg. assistant and docent) were subordinate, and instituted four academic ranks with equal rights and obligations. It also granted to students (who in Greece are organised according to political party ideology) wide participatory and voting rights in faculty meetings, so that in various department assemblies the number of students equals that of teaching staff in voting for chairpersons, deans and rectors. In general, as commonly acknowledged in Greece, university reform has served, in the main, to further politicise the system of higher education and to make an already non-autonomous university more dependent on political factions. In any case, the specific problems of women either seeking or trying to sustain careers in scientific research and university teaching have not been addressed so far, nor have they become issues for public policy.

University professors generally occupy elite positions in the R & D system. To be sure, there are researchers who do not have a university appointment, but directors of research institutes are usually affiliated with a university. More importantly, the director of a research centre is normally a political appointee, and to date has always been a male.

Greek Women in Research: A Pilot Study

While there is no systematic information on women at research institutes, some data is available on women scientists in Greece's most prestigious research centre, the National Research

---

1268/82, has had no substantial impact on this practice which is reinforced by the absence in Greece of a formally organised graduate studies programme in practically all fields. One post-graduate degree is granted in Greece, the Doctorate, on the basis of a thesis prepared under the supervision of a senior faculty member. There are no course requirements for this degree, since there are no organised studies. It is not surprising that a Doctorate obtained abroad is preferred when university appointment is at stake.

Center of Physical Sciences (NRCPS) 'Demokritos'. Of 180 scientists with a Ph.D., thirty-two are women. Of these, seventeen or practically 50 per cent work in biology, five in nuclear or solid state physics, five in chemistry, two in isotopes and the rest in health physics.[7] This is considered an impressive figure by many, which underscores the still undeveloped awareness of the problem in Greece.[8]

We distributed a questionnaire to thirty women scientists working in Demokritos in an attempt to get their own views on two sorts of issues: the general state of science and R & D policy in Greece, and objective obstacles for women choosing careers in science; and personal perceptions regarding their own careers and the potential for Greek women in general. Fifteen women responded in time for the preparation of this study; many, however, communicated personally with the author to express their interest in this research which, as far as they knew, was the first of its kind in Greece. Thus, while the sample is limited, it is not without value as a first general picture of women research scientists in Greece.

The median age of the respondents is about fifty years. The majority are married with two or three children; one is divorced and one is widowed. Except for two, all have received their Doctorate abroad and their positions range from Research Director (Main Investigator) to Senior Research Scientist and to Research Scientist. Their fields of research are biology, chemistry and physics; their salaries commensurate to and in some cases exceed those of university professors. With few exceptions, these women have published articles in international journals.

Most questions called for comments and it is the latter which will be summarised here. With respect to current science policy, almost all respondents cited as problems the lack of adequate funding for both on-going research projects and the introduction of new research. The majority located the government as

7. Our figures here were provided by Dr. Christina Zioudrou, Bio-Chemical Division, Demokritos, in a personal communication, January, 1989.
8. A recent survey of research labour power in Greece (1987) conducted by the General Secretariat of Research and Technology, the main co-ordinator of R & D in Greece, provides figures on all researchers with no breakdown of research staff by sex. The survey, the first of its kind in Greece, is significant nonetheless since it suggests that compared with other EEC countries and Japan, Greece has a very low research force: only two persons per thousand active population are engaged in scientific research.

the central organiser of science policy, a few included the scientific community as well. Some respondents cited as determinants of science policy, government appointed directors of research institutes, and one respondent used the word 'whims' (of directors). The major source of funding so far is the government and EEC grants. In the opinion of most respondents, 'too few' women are involved in decision-making for science and technology even though, according to most, 'a good number' of women are entering the fields of the 'hard' sciences and technology.

The question on the scientific research needs of Greece evoked the most elaborate commentary; these responses can also be taken as some indication of the research interests and needs of these women scientists. Most indicated that government centralised science policy has so far been unresponsive to these needs. The recurring themes include: environmental and ecological research, food and agricultural technological research, raw materials research, cost/effect research, renovation of scientific equipment, hiring of new scientists, and for some respondents, a pressing need is the establishing of a scientific tradition in Greece. Most respondents ascertained that systematic scientific research, dissemination and technical applications (S & T) are pursued largely in research institutes; some included university centres as well. Only two cited the university as the main place of S & T; these were the same two who listed teaching in addition to their independent research work.

Responses to the questions on objective obstacles Greek women face in choosing careers in science and technology overwhelmingly indicated three major factors: educational practices: they are 'not exactly encouraging'; employment opportunities: 'men are preferred' and hierarchy: 'women are on lower rungs'. Among the major subjective obstacles, all respondents cited family or social attitudes and general socialisation.

Most respondents characterised their own careers as 'difficult but rewarding', some as 'full of obstacles', none as 'easy, straightforward'. Almost all would choose the same discipline and career field even though they acknowledge that their male colleagues have an 'easier' career development: 'more is always expected of a woman', 'women have to be doubly qualified and twice as smart', 'men advance at a faster rate than women'. While many acknowledge the help of fathers, mothers and husbands towards their career development as well as 'good teachers' who exerted a positive influence, almost all cited as a major difficulty for career advancement the 'double-burden' of private and public responsibilities which most women must assume.

But the key factors for the promotion of women's entry into the demanding fields of science and technology in Greece, in the opinion of most respondents, seem to be: (1) better education on all levels (the main factor), (2) change in social attitudes and increased respect for woman's work and responsibilities, (3) social programmes to liberate women's time, and (4) actual participation of women in decision-making on science and technology policies.

Some Conclusions and Problematics

Our limited pilot study suggests that women-in-science *qua* problem is intimately linked to women-in-education and women-in-politics. This is not a new idea; in fact it is by now old enough to be taken seriously. The problem of 'why so few?' in science and technology is coextensive with the same query in politics; it has to do with the absence of women in centres of power. In Greece, the major obstacles to women's entry into careers of science and technology include the unexamined views of the function of science itself, and the role of the university in developing social awareness and scientific competence.

With respect to current R & D policy, it is true to say that it is still in the making and still highly charged by political considerations not always compatible with the development of uniform norms and values for science and technology. Greece's minimal funding of research, paralleled only by Portugal among EEC countries, as well as the absence of an effective transmission system among research institutes (these real weaknesses have been noted in the recently formulated Five-Year Plan for R & D policy (GSRT, *Information Bulletin*, 6/89)) operate as inhibiting factors for women who wish to undertake a career in science. For, so far, the lacunae created by some of these peculiarities in the Greek science enterprise have been 'filled', so to speak, by political operators whose *ad hoc* variability compounds rather than resolves the problems. The intrusion of politics has effectively prevented the formation of a scientific community in the strict sense and thus an appropriate forum for deliberation on the real research needs of Greece. This has resulted, not accidently, in the exclusion of many types of research needs and interests as most recently articulated in our pilot study. To be sure, decisions are made, and there is a flourishing of scientific and technological endeavour in Greece, but this is not 'politics-free'. For women, generally not competitive players in the political game, there is an additional burden of having to seek acceptance by largely male-composed politically powerful groups. Thus, Greek women embarking on careers in science encounter not simply male cognitive authority, widely analysed in the feminist literature, but male political authority as well. The latter in Greece has rarely been cognisant or respectful of pluralistic scenarios of the role and function of science.

The recent increased awareness of a university system in disarray, the open dialogue and debate on the aims of university education, are hopeful signs that scientific dialogue will develop enough to launch a substantial review of the real scientific needs of Greece. This is the aim of the Five-Year Plan; it is also, in a very real sense, a condition for the entry of more women into careers of science and technology. Not a sufficient condition, to be sure, but a necessary one.

References

CACOULLOS, A., *Gynaikes kai Nea Technologia sten Ellada* (Women and New Technology in Greece). Athens: ETEBA, 1988.

ELIOU, M., 'Those Whom Reform Forgot', *Comparative Education Review*, vol. 22, no 1, 1978, pp. 60–70.

——, 'Gynaikes panepistemiakoi: exelixi tes theses tous e stasimoteta?' (University women: development or immobility?). *The Greek Review of Social Research*, 70, 1988, pp. 3–24.

FRANGOUDAKI, A., *Koinoniologia tes Ekpaidevsis*. (Sociology of Education). Athens: Papazeses, 1985.

Greek Secretariat for Research and Technology (GSRT), Ministry of Industry, Energy and Technology: *Information Bulletin*, 6, 1989.

LAMBIRI-DIMAKI, J., 'The changing position of women in Greece', in Z. Tzannatos (ed.), *Socialism in Greece: The first four years*. Aldershot: Gower, 1986, pp. 99–107.

8

Women in Academic Science Careers in Turkey

Feride Acar

Academic Women in Turkey: Some General Observations

As in the case of most professions, women in Turkey are not excluded from the academic community of science. Of the approximately 30,000 faculty members in Turkish universities more than 9,000 (32.2 per cent) are female. In view of the fact that the overall literacy rate for Turkish women was 65 per cent (as opposed to 86 per cent for men) in 1985, this phenomenon is clearly impressive. Moreover, historical trends show that women's participation in most scientific fields in the academic world – despite occasional fluctuations – has been increasing since the 1940s.[1]

In such areas as the natural sciences, medicine and even engineering, where women are generally under-represented academically in the western industrialised countries, in Turkey they comprise impressive percentages of the total. For instance, currently (1989), about 32 per cent of the academic personnel in natural sciences, 35 per cent in medicine and health-related fields, and 24 per cent in engineering are females.

Although women's participation rates in some fields (i.e. humanities, fine and applied arts, and medicine) are above the overall participation rate in all fields, in Turkey academic women are not exclusively concentrated in fields generally considered appropriate for 'feminine' identity. On the contrary, particularly in the earlier years of the Republic, the proportion of

---

1. Women were admitted to the academic professions for the first time in 1932–3, but their larger-scale recruitment started in the 1940s (Köker, 1988).

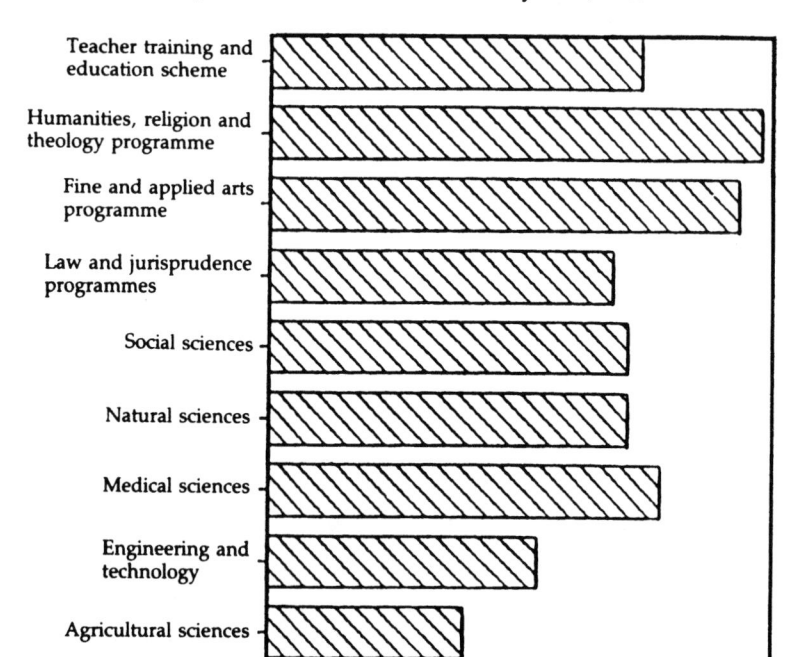

Figure 8.1: Proportion of women academics by field, 1989

[Bar chart showing percentages for:
- Teacher training and education scheme
- Humanities, religion and theology programme
- Fine and applied arts programme
- Law and jurisprudence programmes
- Social sciences
- Natural sciences
- Medical sciences
- Engineering and technology
- Agricultural sciences
- Other fields
- Total of women academics

X-axis: 0, 6, 10, 16, 20, 26, 30, 36, 40, 45 %]

women in natural sciences and engineering departments was higher than in the social sciences, humanities or education. Currently, they are more or less evenly distributed among fields of science (see Figure 8.1).

The number and share of women in all academic positions have also increased with time in Turkey. (Köker, 1988: 296) Today, 20 per cent of all full professors, 23 per cent of all associate professors, and 27 per cent of all assistant professors are women.

As is often found, the proportion of women in lower academic positions is higher than their overall participation rate. In 1988–9, while 32.2 per cent of all academic personnel were

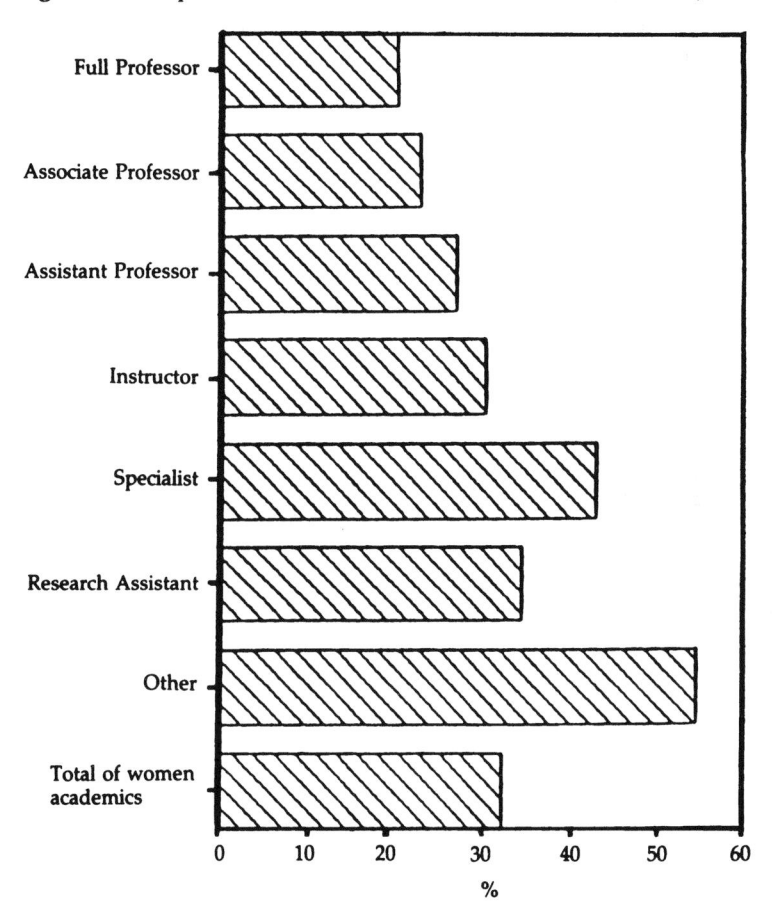

Figure 8.2: Proportion of women with different academic status, 1989

females, 34.6 per cent of all research assistants, 43.0 per cent of all specialists, and 53.4 per cent of all 'other' personnel – which is mainly made up of language instructors – were women (see Figure 8.2).

Like the figures on women's distribution in various fields generally, these rates have also been interpreted as signs of women's improved and enviable position in the academic institutions and the scientific community in Turkey. Supporting these claims is the fact that in Turkish academia women appear to be conspicuously free from traditional constraints placed on women by the values and practices of most traditional societies.

Not only have some women occupied the most important and prestigious positions in the university system (including those of university president and member of the Higher Education Council), but studies show that mostly academic women themselves do not suffer from discrimination and/or institutional obstacles in their career paths. (Acar, 1983; Köker, 1988)

Some Historical and Sociological Factors Underlying Women's Participation in Academia

While these fairly encouraging signs observed in women's situation in the Turkish academic community are not altogether unique to this country (see for example, Ruivo, 1987:388), their specific evolution is best explained by reference to the political, social, and economic forces at work in the last sixty years of Turkish history.

After the First World War, following the disintegration of the Ottoman Empire, the newly established state of the Turkish Republic under the leadership of Mustafa Kemal Atatürk put into action a series of reforms. Among those, reforms improving the social and political conditions of women in Turkish society received a high priority. The modernising elite of the Republic perceived women's higher education and career acquisition as a sign of Westernisation – a professed goal of the Republic. Consequently, state ideology and the elite subculture which it defined functioned in such a way that they encouraged and supported women's careers as part of their 'development' mission.

The newly structured universities designed to reflect the image of 'modern' Turkey emerged as particularly suitable media for operationalising the ideals of the Republic regarding women. Under these circumstances daughters of middle- and upper-class urban elite families who identified with Kemalist ideals benefited not only from the material advantages of their social background – which made it possible for them to receive advanced education and often facilitated their pursuit of a career by enabling them to hire household and child care help (Öncü, 1981; Erkut, 1982) – but also from the supportive and legitimising values of the elite subculture in which they were raised.[2] It is

---

2. It also has been argued that, in Turkey, it was possible for women of elite

not surprising, therefore, that women interviewed in two independent studies done on academic women in Turkey invariably reported being highly motivated in their career goals and attributed this to their family socialisation and particularly to the influence of their fathers. (Acar, 1983; Köker, 1988)

Furthermore, women's rather unusual presence in the 'non-conventional' fields can also be interpreted in the light of this elite culture. Republican state ideology was positivist; it glorified 'hard' sciences *vis-à-vis* humanities and social sciences. Women socialised in the elite subculture were thus deliberately oriented towards careers in such fields as natural sciences and mathematics by the dominant ideological discourse. Hence, in Turkey we observe a non-conventional distribution of women among different fields in academia, particularly in the earlier years. For instance, in 1946–7, 44 per cent of the academic staff of the Faculty of Natural Sciences were female. In the same year only 22 per cent of the humanities faculty were women. (Köker, 1988) It is only after the decline in the effectiveness of the Kemalist message and political ethos in the post-1950 period that women's participation in the 'non-conventional' area of natural sciences has decreased noticeably. Nevertheless, the relative absence of sex-typing of academic fields in Turkish society is an obvious outcome of this heritage. (Öncü, 1981)

It is possible that the favourable milieu women generally found in the universities in Turkey is also partly a consequence of the fact that, as in other developing countries, academic science has always had rather weak links with real power centres in society. (Ruivo, 1987) Thus, at the risk of sounding conspirational, one can speculate that despite the ideological rhetoric placed on science and scientists in Turkey, they have not been crucial enough to merit 'protection' from women. It is noteworthy in this sense that in the early years, when women were being especially encouraged and supported to enter 'science' fields, they remained conspicuously absent among

---

backgrounds to fill prestigious positions in society during rapid economic and social change not only because they were available and willing, but also because political authorities acting with social class bias preferred women from the upper classes to men from lower-class origins (Öncü, 1981). Thus, it can be assumed that universities in the early Republican era opened their doors to women as much out of the need to fill the newly created positions as out of eagerness to demonstrate their loyalty to the ideals of modernisation.

academics in faculties of law and political science (Köker, 1988) – institutions most clearly associated with the state and wielding of political power in Turkey.

An Analysis of Women's Participation in Present-Day Academia

Against this background, questions such as 'What does women's present situation in academia imply beyond the obvious signs of relative improvement?' and 'How do women really fare vis-à-vis men after sixty-five years of official egalitarian ideology and practice?' are crucial in the Turkish context. The following analysis based on a combination of the results of macro-level hard data and interpretation of micro-level qualitative information attempts to provide some answers.[3] Figures presented in the text summarise the findings; more detailed data are presented in Tables 8.1–8.5.

Women and Academic Status

As pointed out above, the share of women in academic science has continuously increased over the years to reach impressive overall proportions in 1989. The proportion of women in Turkish academia has increased from 18.9 per cent in 1960, to 25.4 per cent in 1980, and to 32.2 per cent in 1989. Although this increase has been reflected in all academic positions, women in academic science still continue to have higher representation in lower echelon positions. The ever increasing participation of women over a long period of time in academia has failed to bring them to an equal footing with men in promotion. As can be observed in Figure 8.2, women's share is lower in the more senior academic posts. Similarly, only 8.7 per cent of all women are full professors (as opposed to 16.6 per cent of all men) in the

3. The 1989 statistics on which the present analysis is based were obtained from the extensive database compiled for the Higher Education Council and made available to the author by the Student Selection and Placement Centre. The qualitative data are based on the findings of two studies, Acar, 1983 and Köker, 1988, both of which collected information on the social background characteristics, role priorities, career motivations, commitments, role perceptions, performances and attitudes of two small samples of academic women through semi-structured in-depth interviews.

Table 8.1: Average time spent between two consecutive degrees/titles (years)

	B.S. to M.S.	M.S. to Ph.D.	Ph.D. to Assoc. Prof.	Assoc. Prof. to Full Prof.
Women	4.2	6.1	6.7	7.7
Men	4.2	6.0	6.2	7.5

Source: Constructed from data compiled by the Student Selection and Placement Centre

academic world. Parallel differences between men and women exist across other academic status categories. For instance, 4.8 per cent of academic women are associate professors (vs. 7.8 per cent of men), and 9.5 per cent are assistant professors (vs. 12.3 per cent of men).

Two factors may be thought to contribute to the present concentration of women in the lower echelons. One of them is the entry of more women into the academic world than men in the recent decades; this is readily observed in the increasing share of women throughout the past decades. An alternative explanation may be sought in the slower promotion rates of women. However, our data show that the average time spent between degrees is not considerably different for women and men academics.

Although Table 8.1 shows a trend favouring men, it falls short of accounting for the present skewed distribution. Hence, rapid entry of women appears as the more plausible factor contributing to lower echelon concentration, and in conjunction with the narrowing gap between men and women even in the higher echelons (Köker, 1988: 302), it can be interpreted as an encouraging sign for the future. However, close to one-third of the women in the academic world are instructors (14.1 per cent of all women) or language instructors (14.8 per cent of all women).[4] Since these lower echelon positions are to a large extent 'dead-end' career tracks, the optimistic message of the increasing overall women's participation in the academic profession should be regarded with some caution.

4. In Figures 8.2 and 8.3, the category 'other' is mainly made up of language instructors.

Literature suggests that women all over have problems with careers due to the presence of psychological, sociological and institutional barriers. (Stolte-Heiskanen, 1988) It is further suggested that women who are not barred or discouraged from entering scientific professions in the beginning are often weeded out before they reach higher or key posts. While specific historical, economic and political forces may facilitate women's entry into professional careers, their elimination from the race at later points is often attributed to a medley of factors influencing women's lives.

The discrimination mechanisms based on patriarchal values in society and the institution of science; social and cultural pressures causing role incompatibility between careers and family obligations; and psychological factors influencing women's own self-images are usually listed as jointly causing obstacles to the development of women's careers.

In the Turkish case, qualitative data obtained from women academics supports these hypotheses to varying degrees. Generally, Turkish academic women report having received fair and equal treatment in the academic world, their perceptions thus supporting the optimistic interpretation of the hard data. So far as most women are concerned, discrimination in the university has simply not been their personal experience. (Acar, 1983; Köker, 1988) However, it has been argued elsewhere (Acar, 1983) that Turkish academic women's conception of equality was 'formalistic'. Thus, only in cases where women conceived of gender equality in more than 'formal' terms, could they identify subtle reflections of discrimination as forces acting as barriers to their career advancement. In fact, it was found that such women often complained of exclusion from informal collegiate networks that led to their being less informed and less influential in the academic institutions' internal politics. (Acar, 1983)

While the presence or importance of outright discrimination in academic institutions remains a controversial point, interview data on Turkish academic women clearly reveal that women's careers in academia are seriously affected by role conflict. Role conflict becomes increasingly more important in the lives of academic women as their careers progress. It is often reported to be the cause of many women's redefinition of their career roles. For instance, women who were interviewed in one study, stated that pressures rising from the increasing demands of their family roles often caused them to reduce their standards of performance in their careers even though they had received very supportive family socialisation and experienced very high career motivation at the beginning. (Acar, 1983)

Consequently, it may be difficult to explain the disproportionately higher representation of academic women in lower echelon positions with the existence of discriminatory practices of the academic institutions in Turkey. It is, however, clear that role conflict resulting mainly from the inability of women scientists to effectively dissociate themselves from traditional family roles – despite their social background and socialisation characteristics – continues to be an important barrier to women's career advancement in academia.

Women in Different Fields

Figure 8.1 also provides data on the distribution of women into various science fields. It shows that currently women have their highest share in the humanities (43 per cent) and fine and applied arts (41 per cent).

Compared with women's shares among the faculty in these and other fields in the past, this situation reflects a change. Women's shares have increased in all fields except natural sciences over the years. However, they have not done so at similar rates in all fields. Compared to their shares in these fields in 1981–2, the increases in women's participation in the humanities (gain of 23.1 percentage points) and law (gain of 14.6 percentage points) have been most noticeable against an overall increase of 6.8 percentage points in all fields. The increase in women's share has been minimal in medicine (gain of 1.2 percentage points) because the share in 1981–2 was already among the highest, and in the natural sciences (gain of 3.1 percentage points) where women's share had already dropped significantly from what it was in the early years of the Republic.

Sociological factors may be thought to explain this change. One such influence undoubtedly is the fact that the effectiveness of early state policies has somewhat worn off in society. Thus the ideological mobilisation of elite women into natural science disciplines lost its impetus.

Also in the post-1950 period, increased social mobility in the country has meant more competition for elite women from the upwardly mobile males of lower socio-economic origins. Consequently, it is possible to infer that women lost their 'preferred' status of entry into the academic professions. Most women coming into the system in the 1960s and 1970s were less ideologically motivated, some were more influenced by the 'traditional' values of society, many were more likely to choose a field that is associated with a well-paying profession, and all had to face tougher competition from men. In this period, while expansion of educational facilities to larger segments of the population has meant more and more women (as well as men) opting for an academic career, women's choice of fields shifted from the 'unconventional' to the conventionally more 'feminine' fields. All these factors have resulted in a distribution of Turkish women academics in science fields that is more similar now to the typical Western model.

In terms of the distribution of women academics in different fields, a point that warrants attention concerns women's participation in the sub-specialities of certain fields. In the Turkish case, qualitative data suggests that women's participation tends to decrease as a sub-field becomes more competitive. For instance, it has been observed that in medicine and health-related sciences the women's share is higher in pharmaceutical science and dentistry as opposed to medicine *per se*. By the same token among the sub-fields of medicine, women have been found to constitute larger shares in paediatrics but distinctly smaller ones in surgery. (Köker, 1988)

Fields in which the highest proportions of women academics are senior faculty[5] are natural and social sciences, medicine and law. Among these, in medicine where of all women faculty 15.3 per cent are professors and 6.6 per cent are associate professors, women appear to fare the best. It is interesting that law, where women's entry into the field was more recent than in others, constitutes the second best field in terms of proportion of academics promoted to full professor (12.7 per cent of all women faculty are professors). Paradoxically, in the humanities where women's share increased most impressively in the 1980s, senior

---

5. For purposes of this analysis 'senior faculty' has been defined as full and associate professors; the two highest academic positions whose members have tenure.

Table 8.2: Distribution and share of women academics in different academic positions by fields, 1989

	1	2	3	4	5	6	7	8	9
Teacher Training and Education Science	34 (3.3)	30 (2.9)	78 (7.6)	488 (47.6)	20 (2.0)	178 (17.5)	194 (19.0)	1 020 (100.0)	– (32.6)
Humanities, Religion and Theology Programmes	66 (5.0)	29 (2.2)	49 (3.7)	170 (12.8)	27 (2.0)	151 (11.3)	841 (63.1)	1 333 (100.0)	– (43.9)
Fine and Applied Arts Programmes	13 (2.9)	31 (6.9)	45 (10.0)	233 (52.0)	3 (0.7)	78 (17.4)	45 (10.0)	448 (100.0)	– (41.7)
Law and Jurisprudence Programmes	14 (12.7)	5 (4.5)	11 (10.0)	7 (6.4)	3 (2.7)	70 (63.6)	0 (0.0)	110 (100.0)	– (30.2)
Social Science (Social, Behavioral, Commercial)	70 (8.0)	53 (6.0)	132 (15.0)	94 (10.7)	38 (4.3)	457 (51.9)	36 (4.1)	880 (100.0)	– (31.5)
Natural Sciences	100 (9.5)	60 (5.7)	158 (15.0)	64 (6.1)	51 (4.8)	617 (58.7)	2 (0.2)	1 052 (100.0)	– (31.6)
Medical Sciences	456 (15.3)	195 (6.8)	277 (9.3)	131 (4.4)	114 (3.8)	1 793 (60.2)	13 (0.4)	2 980 (100.0)	– (34.6)
Engineering and Technology	64 (5.2)	57 (4.6)	139 (11.3)	126 (10.2)	69 (5.6)	740 (60.2)	35 (2.8)	1 230 (100.0)	– (23.6)
Agricultural Sciences	34 (10.8)	10 (3.2)	30 (9.5)	7 (2.2)	17 (5.4)	210 (66.7)	7 (2.2)	315 (100.0)	– (17.4)
Other Fields	6 (1.4)	2 (0.5)	12 (2.7)	61 (13.8)	45 (10.2)	34 (7.7)	282 (83.8)	442 (100.0)	– (38.4)
Total number of women academics	857	473	931	1 379	387	4 328	1 455	9 810	(32.2)
Total number of academics	4 284	2 086	3 480	4 552	900	12 521	2 674	30 497	–
Ratio of women to total academics	(20.0)	(22.7)	(26.8)	(30.3)	(43.0)	(34.8)	(54.4)	(32.2)	–

Source: Constructed from data compiled by the Student Selection and Placement Centre of Turkey.

The figures in parentheses indicate percentages (the proportion of women in a given position compared to the total number of women academics in that field)

1 = Full Professor; 2 = Associate Professor; 3 = Assistant Professor; 4 = Instructor; 5 = Specialist; 6 = Research Assistant; 7 = Other; 8 = Total Women Academics; 9 = Overall percentage of women.

Figure 8.3: Distribution of women into academic status categories by field, 1989

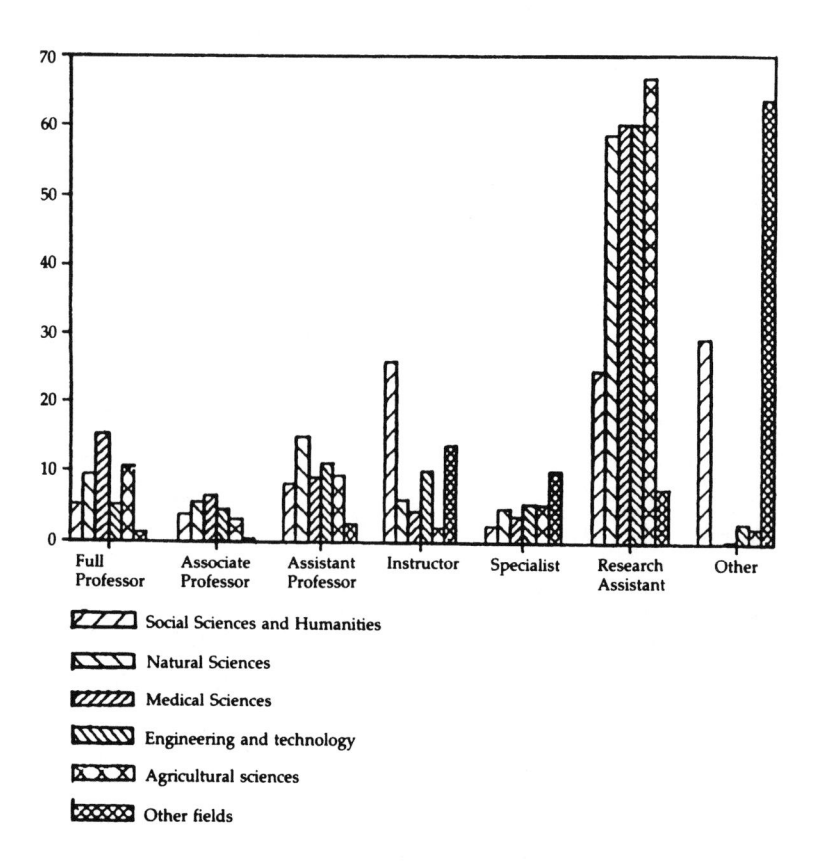

[Bar chart showing distribution across categories: Full Professor, Associate Professor, Assistant Professor, Instructor, Specialist, Research Assistant, Other, with legend indicating: Social Sciences and Humanities, Natural Sciences, Medical Sciences, Engineering and technology, Agricultural sciences, Other fields]

women faculty are relatively few. In fact, women even fare better in engineering and agricultural fields where their overall shares are smaller.

Geographical Distribution of Women Academics

Turkey has thirty universities – one private, the rest public – and fifteen of these are located in the largest urban centres of Istanbul, Ankara and Izmir. The universities in these three metropolitan centres employ 65.6 per cent of all faculty, and 76.6 per cent of all women faculty. Figure 8.4 presents a distribution of

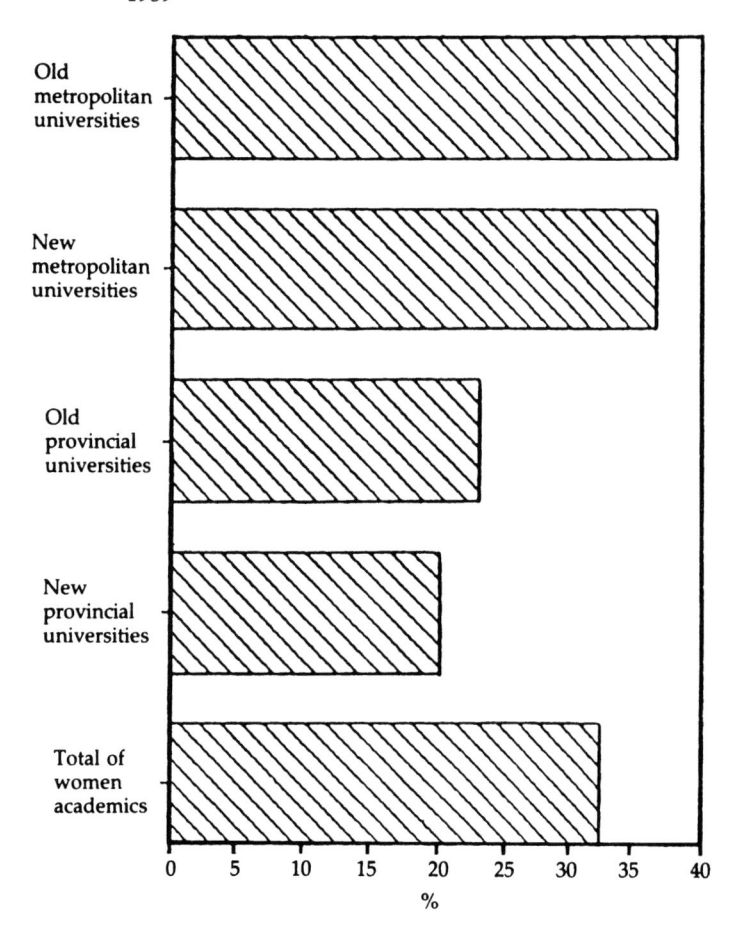

Figure 8.4: Proportion of women academics by type of university, 1989

[Bar chart showing percentages for: Old metropolitan universities (~38%), New metropolitan universities (~36%), Old provincial universities (~23%), New provincial universities (~20%), Total of women academics (~32%)]

women faculty in four groups of universities.[6]

Women faculty constitute increasingly larger portions of all faculty as one moves from new provincial universities to old metropolitan ones. Of the total number of women academics in Turkey, more than half (53.8 per cent) work in old metropolitan

---

6. The metropolitan/provincial distinction refers to the institution's location in one of the three big cities vs. other cities in Turkey. The old–new axis refers to its date of establishment. Those universities which existed before 1981, when a major reorganisation of the Turkish higher education system was carried out, are marked as old; others as new.

universities; only 7.6 per cent in the new provincial universities. It is noteworthy that in the older provincial universities located in mid-sized urban centres throughout Turkey where one-fourth of all male academics are based, only 15.8 per cent of the women can be found.

Interview data from women in old metropolitan universities have revealed that academic women's orientation towards scientific careers and their subsequent career performance was almost always due to the family socialisation and support they had received. (Acar, 1983) While such supportive attitudes and behaviour are often enjoyed by daughters of urban educated parents of the middle and upper class in Turkey, they are, to this day, less common in families of lower socio-economic background or provincial tradition. Furthermore, career women in general have more limited geographical mobility than men, and women of privileged backgrounds are especially unlikely to change their locations in order to accept appointments in the new universities located in the economically less developed and culturally more traditional regions of the country. Thus, the bases of female recruitment are small for new provincial universities. Moreover, as one goes from new provincial universities to the old metropolitan ones, both the overall share and the proportion of women as full, associate, and assistant professors increases. That is to say, women faculty in old metropolitan universities are more likely to have higher posts than women in other institutions.

Since older metropolitan universities generally rank higher in the nation's scientific hierarchy and are more active and influential centres of research, women's concentration in relatively higher positions in these institutions has a positive side to it. On the other hand, the opening of new universities in provincial settings has so far not given any signs of a potential to offset the existing inequality in women's conditions in Turkish academia. Explanations of this phenomenon can be found in regional and rural–urban differences in modernity in Turkey.

Women in Academic Administration

Administrative positions in the Turkish universities – despite the additional non-academic loads they imply – are often wanted and sought after by faculty members. Within the

Table 8.3: Distribution and share of women academics in different academic positions by type of university, 1989

	1	2	3	4	5	6	7	8	9
Old Metropolitan Universities	665 (77.6)	267 (56.4)	469 (50.4)	499 (36.2)	202 (52.2)	2 432 (56.2)	747 (51.3)	5 281 (53.8)	13 879 (38.1)
New Metropolitan Universities	66 (10.3)	122 (25.8)	240 (25.8)	539 (39.1)	58 (15.0)	812 (18.8)	377 (25.9)	2 236 (22.8)	6 124 (36.5)
Old Provincial Universities	88 (7.9)	52 (11.0)	162 (17.4)	250 (18.1)	77 (19.9)	725 (16.8)	218 (14.8)	1 550 (15.8)	6 770 (22.9)
New Provincial Universities	36 (4.2)	32 (6.8)	60 (6.4)	91 (6.6)	50 (12.9)	359 (8.3)	115 (7.9)	743 (7.6)	3 724 (20.0)
Total number of academics	4 284	2 085	3 480	4 552	900	12 521	2 674		
Ratio of women to total academics	(20.0)	(22.7)	(26.8)	(30.3)	(43.0)	(34.8)	(54.4)		

Source: Constructed from data compiled by the Student Selection and Placement Centre of Turkey.

The figures in parentheses indicate percentages (the proportion of women in a given university compared to the total number of women in that academic position).

1 = Full Professor; 2 = Associate Professor; 3 = Assistant Professor; 4 = Instructor; 5 = Specialist; 6 = Research Assistant; 7 = Other; 8 = Total Women Academics; 9 = Ratio of Women to Total Academics.

Table 8.4: Distribution and share of women academics in administrative assignments by type of university, 1989

	Women Academics Holding Administrative Assignment		
	Higher level	Lower level	Total
Old Metropolitan Universities	31 (19.6)	334 (22.1)	365 (21.9)
New Metropolitan Universities	4 (4.3)	138 (17.6)	142 (16.2)
Old Provincial Universities	9 (6.9)	108 (11.0)	117 (10.5)
New Provincial Universities	3 (3.0)	58 (10.4)	61 (9.3)
All Universities	47 (9.8)	638 (16.7)	685 (15.9)

Source: Constructed from data compiled by the Student Selection and Placement Centre of Turkey.

The figures in parentheses indicate the proportion of women compared to total appointees.

academic institutions, they are evaluated as seats of power, and outside they are considered prestigious. Particularly top-level administrative positions, such as university president or faculty dean, not only provide individuals with the ability to exercise considerable power within their organisations, but also they often bring recognition and respectability on the local or even national scale.

In 1989, only 15.9 per cent of all administrative appointments in Turkish universities were held by women (see Table 8.4). As can be expected, women constitute bigger shares of administrators in old metropolitan universities (21.9 per cent) and their percentages among administrators decline as one moves to new and/or provincial universities (16.2 per cent, 10.5 per cent and 9.3 per cent).

In all four groups of universities women constitute bigger percentages among holders of lower as opposed to higher level administrative posts.[7] However, while differences are pro-

---

7. Lower level administrative posts include department chairs, assistant

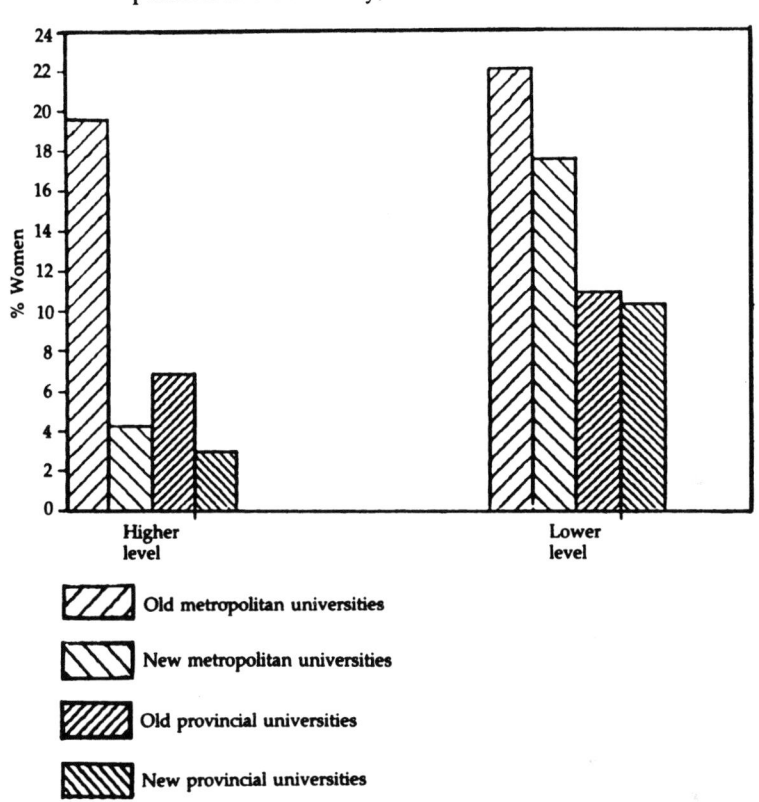

Figure 8.5: Share of women in administrative positions by type of position and university, 1989

- Old metropolitan universities
- New metropolitan universities
- Old provincial universities
- New provincial universities

nounced in the other three categories, in old metropolitan universities where women hold 19.6 per cent of all higher level administrative positions, the percentages of women administrators in higher and lower level offices are very close (19.6 per cent vs. 22.1 per cent). In other words, while women on the whole are under-represented among administrators in Turkish universities and are often confined to lower level positions, in the well-established metropolitan universities they once again

---

chairs, other intra-department appointments, computer centre directors, etc. Higher level administrative posts include university presidents and vice-presidents, deans and vice-deans of graduate and undergraduate faculties, directors and assistant directors of research institutes.

fare better than in other institutions. Obviously, this is due mostly to the bigger presence of women among the senior faculty (full and associate professors) in these universities.

Our data also suggest that while the chances of holding administrative office decline for everyone regardless of sex, as one moves from full professors to holders of lower academic positions, the lower the academic position the less likely is the appointment of a female to an administrative office. In other words, sex plays a bigger role in women's access to institutional power when women are junior faculty; it declines in importance with academic promotions (and concomitantly age).

Since universities located in metropolitan and provincial settings often operate in significantly different social and cultural milieus, the above-mentioned relationship between gender, academic rank and administrative power also varies with respect to the type of university. Thus in new provincial universities the younger and more junior women faculty are more disadvantaged in their competition with men of similar status for administrative office than are senior female faculty in the well-established universities in Istanbul, Ankara or Izmir.

This situation may be explained by a variety of factors. It is expected that patriarchal values would be more effective in provincial 'traditional' environments and with regard to younger women of junior status. In these settings women are more likely to be discriminated against. They themselves are also more likely to have internalised traditional patriarchal norms defining women's primary role as wife and mother, leading to voluntary withdrawal from competition with men.

In the Turkish case, interview data show that women in the old metropolitan universities express very high confidence in their ability to hold different types of administrative office.[8] In fact, many believe that women are better suited for demands of administrative positions because they are more 'flexible', 'con-

---

8. This situation is in distinct contrast to the attitudes observed among academic women in more traditional social environments. For instance, a study comparing a sample of Turkish academic women from the Middle East Technical University and Ankara University – two institutions falling into the category of old, metropolitan universities in Turkey – with academic women from Yarmouk University in Jordan indicated that academic women in the latter case, to a larger extent, considered administration to be 'men's work' and wanted no part of it; contrary to Turkish academic women, they usually perceived themselves to be out of the main race in academia (Acar, 1986).

genial' and 'hard working' (Acar, 1983; Köker, 1988) than men. These women usually point only to the presence of an overload of demands placed on them by ever increasing obligations of their family roles as the reason for their withdrawal from the competition for administrative positions. Occasionally, some also attribute a role to the chauvinistic attitudes of male colleagues. However, the discrimination argument was only stressed by a few younger women in these studies. (Acar, 1983; Köker, 1988)

According to the majority of academic women in the studies reported here, women who shy away from administrative positions in Turkish universities do so on account of role conflict rather than institutional discrimination. As will be elaborated in the next section, it is possible to provide additional support – however circumstantial – for the above argument from existing statistics on administrative appointment and marital status.

All in all, with respect to women's access to the seats of power in academia, different types of data appear to converge on a few points. Senior academic women in metropolitan universities of established tradition and 'modernist' cultural heritage, although still a minority among academic administrators, are not especially disadvantaged *vis-à-vis* men. These women are found to express high self-confidence and self-esteem as well as a competitive attitude with regard to their careers. (Acar, 1983) Their psychological profiles also reflect an apparent immunity[9] to social-psychological pressures associated with the 'fear of success' observed in achievement-oriented females in Anglo-Saxon cultures.

To the extent that these women are pioneers and precursors of the future and they function as role models in society, their position in the hierarchies of the more prestigious academic institutions is an encouraging sign. However, the considerable existing differences between women's conditions in different types of universities and the known historical and socio-cultural factors that explain these differences render it difficult to make generalisations or predictions for all women in Turkish academia on the basis of the characteristics, experience and current positions of senior academic women in metropolitan universities.

9. The culture-specific nature of 'fear of success' has been demonstrated by a study of female university students in Turkey who are reported not to exhibit such attitudes. (Kandiyoti, 1981)

Academic Women and Marriage

Statistics on marital status of male and female members of Turkish academia exhibit noticeable differences. Currently 45 per cent of all female academics as opposed to 36.2 per cent of all male academics are single. Moreover, 3.9 per cent of all women academics versus 1.9 per cent of men are either divorced or widowed. Thus, it is interesting to observe that about half (48.9 per cent) of all women academics are unmarried whereas the corresponding figure (38.1 per cent) for men is significantly lower.

This difference is partly due to the fact that women faculty have a larger share of younger members. However, another possible interpretation is that women have more difficulty than men in combining an academic career with marriage. There is empirical support for this factor across categories of academic status. For instance, 15 per cent of male assistant professors vs. 25.1 per cent of female assistant professors are single. The difference is even more pronounced in the case of senior faculty: 22 per cent of female full professors are single as compared to only 4 per cent with men. These figures indicate that senior female faculty are more likely to be single than the junior women faculty, and much more so than their male counterparts. Thus, being single and women's advancement in the academic profession are clearly related.

This relationship is further strengthened by the data on divorced and widowed women faculty. While due to the nature of the database, it is not possible in this study to distinguish the exact percentage of the widowed – an obviously involuntary status – from divorced, it can be assumed that regardless of the cause of their current status, Turkish academic women, in general, are more likely to stay unmarried than academic men. While 1.8 per cent of male full professors are divorced or widowed, 9.0 per cent of all female full professors fall into this category. Successful women in the academic world seem to find it more difficult to pursue their careers within a marriage than their male counterparts do.

Table 8.5 also provides evidence on the relationship between academic women's marital status and the likelihood of being in administrative office. Analysis of figures here shows that a very high percentage of women who currently have administrative

Table 8.5: Distribution and share of academics holding administrative appointment by marital status, 1989

Marital Status	Higher level Women	Higher level Men	Administrative Appointment Lower level Women	Lower level Men	All levels Women	All levels Men
Married	30 (63.8)	441 (94.9)	459 (71.9)	2 963 (92.8)	489 (71.4)	3 374 (93.1)
Single	14 (29.8)	13 (3.0)	129 (20.2)	182 (5.7)	143 (20.9)	195 (5.4)
Divorced or Widowed	3 (6.4)	9 (2.1)	50 (7.8)	48 (1.5)	53 (7.7)	57 (1.6)
Total	47 (100.0)	463 (100.0)	638 (100.0)	3 193 (100.0)	665 (100.0)	3 626 (100.0)

Source: Constructed from data compiled by the Student Selection and Placement Centre of Turkey.

The figures in parentheses indicate percentages (the proportion of appointees with the given marital status compared to the total number of appointees in the same column).

positions are unmarried: 20.9 per cent of female administrators are single and 7.7 per cent are divorced/widowed. The corresponding percentages for male administrators are only 5.4 per cent and 1.6 per cent respectively. Divorced or widowed women have a higher propensity (13.9 per cent) to hold office than either single (3.0 per cent) or married (9.8 per cent) women. Moreover, an equally interesting observation in Table 8.5 is that of all the women in high administrative office 29.8 per cent are single, whereas of the men only 3 per cent are single.

These statistical figures and the information gathered through the interviews, once again converge on the observation that for most of the academic women the conflict between family and career roles is crucial as an obstacle to career advancement. While married academic women in Turkey try to cope with this dilemma in a variety of ways (Acar, 1983), figures reported above show that a substantial number avoid such conflict by dissociating themselves from family roles, i.e. by being unmarried.

Although in the interviews women who were single or divorced generally attributed these facts to factors unrelated to their careers, this can be attributed to the widespread prevalence of socio-cultural values that expect satisfactory performance from women in both family and career roles. That is to say, academic women who function in a subculture that expects them to be 'superwomen' were not ready to admit that their 'success' in the career role was contingent upon their 'failure' in their family role, possibly for reasons of social desirability.

A different aspect of the relationship between academic women's career performance and their marital status is the husband's occupation. The women academics in Turkey seem to be firm believers in 'inbreeding': 35 per cent of all the married women in the university system have husbands who are also faculty members.

It is noticeable that women who were interviewed, regardless of whether they were themselves married to colleagues or not, thought professional endogamy was a career-promoting force for women in academia. Those who had academic husbands generally stated that their choice of spouse had indeed facilitated the realisation of their career goals. (Acar, 1983; Köker, 1988) This type of support ties in with the casual observation that for an even higher percentage of senior women faculty the husbands are in academia.

It appears that academic husbands help women's career objectives in different ways. For instance, it was commonly believed that 'being in the same boat' made the husband more understanding of the wife's needs and career role demands in everyday life.

Although most of the Turkish academic women did not 'share' household and child care responsibilities with their husbands on an equal basis, academic husbands who had more flexible time schedules and less 'conservative' attitudes were reported to be more 'helpful' in these matters. (Acar, 1983; Köker, 1988)

In some cases, academic women's marriage to men with parallel career objectives had helped provide early career opportunities for the women. Several women said going abroad for graduate study had been contingent upon their marriage.

It is also likely that to some extent marriage to a fellow academic functions as a guarantee of the seriousness and reliability of the woman's career intentions so far as the academic community is concerned. Thus, it may function as an effective mechanism in beating the odds against women in academic science. Consequently, one can hypothesise that, next to being unmarried, being the wife of a fellow academic is an effective factor promoting women's career advancement in Turkish academia.

Concluding Remarks

The preceding analysis, taking extent of participation, academic status and administrative appointment as indicators, evaluates women's position in academic science as it is influenced by their marital status, the location and type of university they are in, and their field of specialisation.

The findings indicate that one of the salient obstacles to women's career advancement in Turkish academia is the debilitating influence of the conflict in family and career roles. So long as academic women have to seek and find personal solutions to the conflict between their career and family roles by either altogether withdrawing from the family role or by trying to 'shoulder an overload of responsibilities' (Erkut, 1982) – as they currently appear to do – the future does not promise significant change in the less than equal role they play in science and academia. The first of these options (staying unmarried) cannot be a widely practised alternative in any society, and the second (being a 'superwoman') is not only very hard to realise but also takes a tremendous toll on women in terms of physical and psychological adjustment.

Turkish experience in the last sixty years has demonstrated that through deliberate state policies women's entry into the academic community has been made possible. Yet, in the absence of widespread social-structural mechanisms to provide for the redefinition of women's family roles, the situation of women in the academic world is still far from being satisfactory today. The failure to affect such deep-rooted change reflects itself in the recent trend toward concentration of women in the more 'feminine' fields; pseudo-participation in academia in the form of increasing numbers of women in lower level 'dead-end' positions, and less commendable positions of women in the newer universities.

Although the early republican ideological legitimation of policies directed to the improvement of women's positions has suffered lapses and reversals in time, it has created a momentum in the right direction. This momentum still accounts, to a large extent, for the absence of sex-typing of professions and fields in Turkish society. This valuable head-start needs to be utilised as a stepping stone for policies to further women's meaningful participation in science. In this context, women's participation in the new provincial universities acquires a central importance not only for today but also for the future. Today, what looks like the evolution of a dual academic structure is taking place in Turkey with regard to women. On the one side of this duality are old metropolitan universities where women have secured relatively advantageous positions, and on the other are new provincial universities which do not offer opportunities for real participation by women. Measures must be taken to ensure that these new provincial universities do not function to the detriment of women's advancement and equality with men in Turkish academia.

References

ACAR, E., 'Turkish Women in Academia: Roles and Careers,' *METU Studies in Development*, vol. 10, no 4, 1983.
——, 'Working Women in a Changing Society: The Case of Jordanian Academics,' *METU Studies in Development*, vol. 13, nos 3–4, 1986.
ERKUT, S., 'Dualism in Values Towards Education of Turkish Women' in C. Kağitcibasi (ed.), *Sex Roles, Family and Community in Turkey*. Bloomington, Ind.: Indiana University Turkish Studies, 1982, pp. 121–32.
KANDIYOTI, D., 'Dimensions of Psychological Change in Women: An Intergenerational Comparison' in N. Abadan-Unat (ed.), *Women in Turkish Society*. Leiden: E.J. Brill, 1981, pp. 233–58.
KÖKER, E., 'Türkiye'de Kadın, Eğitim ve Siyaset: Yüksek Öğrenim Kurumlarinda Kadinin Durumu Üzerine Bir Inceleme' (Women, Education and Politics in Turkey: A Study of Women's Condition in Institutions of Higher Education), unpublished Ph.D. dissertation, Ankara University, 1988.
ÖNCÜ, A., 'Turkish Women in the Professions: Why So Many?' in N. Abadan-Unat (ed.), *Women in Turkish Society*, pp. 181–92.
RUIVO, B., 'The Intellectual Labor Market in Developed and Developing Countries: Women's Representation in Scientific Research,' *International Journal of Science Education*, vol. 9, no. 3, 1987.

STOLTE-HEISKANEN, V., *Women's Participation in Positions of Responsibility in Careers of Science and Technology: Obstacles and Opportunities.* Department of Sociology and Social Psychology, University of Tampere, Finland, Working Papers No. 26, 1988.

9

Women at the Top in Science and Technology Fields
Profile of Women Academics at Dutch Universities

Esther K. Hicks

Introduction

Although R & D funding in the Netherlands has witnessed a relative increase both in terms of manpower requirements and job availability – particularly in the industrial sector – the number of women occupying decision- and policy-making positions in this sector and at the government research institution level is virtually nil. At the academic level the percentage of women at the top, while rising slightly at the lecturer level in some areas, remains negligible.

The most recent report of the Ministry of Education and Sciences (1989) indicates that women currently comprise no more than 2 per cent of the total number of professorial and 4.5 per cent of lecturer positions at Dutch universities. In science and technology fields (including the medical sciences) they occupy less than 1 per cent of total professorial and less than 6 per cent of lecturer positions. (Beekes, 1988; Leijenaar, 1987; Hicks and Noordenbos, 1990) This situation has not been ameliorated by the various programmes and organisations developed over the last decade.[1] Nevertheless, although the short-term effect of such programmes has been negligible they may, in the

---

1. For example, the 'choose the exact sciences' programme, directed at girls at secondary school level; the development of women's professional organisations; the generation of policy-directed emancipation programmes.

long term, influence both women's career choices and how they view their academic credibility.

While there has been an increase in the number of women enrolled in exact and technical science programmes at the university level and in those hired for research and technical positions, it is not clear that this has affected the number of women holding lecturer, professorial and/or policy- and decision-making positions. Indeed, these numbers continue to be low, and have even decreased in some areas. Paradoxically, the potential women have to participate in and contribute to science and technology R & D at the academic level, is correlated to the number of women already gainfully employed in such fields; the status and rank positions they occupy; and their concomitant involvement (actual and probable) in decision- and policy-making activities. We might well ask ourselves why it is that, although the number of women studying and employed in exact and technical science fields is on the increase, few of these women are represented in the top echelons of their fields, and fewer still are involved in R & D decision- and policy-making activities.

Although the generally late entry of large numbers of qualified women into the labour arena – more specifically into the science and technology labour market – has restricted women in competing for high level decision- and policy-making positions, this is by no means the only impediment they face.

This study constitutes an exploratory approach to the issue of 'why so few' at the top in the exact and technical sciences. The focus is on those women holding top academic positions (i.e. professorial or lecturer positions) at all major universities in the Netherlands.

Research Design

With the aid of university personnel listings, a questionnaire was distributed to all those women meeting the qualifications of this study, i.e. to those women holding (or having held) a lecturer or professorial position in university departments conducting research in the medical sciences (to include biology, bio-medical research and medical technology), mathematics and the natural sciences, engineering and technical sciences (all) and

Table 9.1: Distribution, by field, of the sampled population and responses obtained

Sampled Population:	Biomedical Med. Tech.	Mathematics Exact	Engineering Tech. sci.	Chemistry	Total
UvA	3[a]	1		1[a]	5
VU	2			1	3
TU Delft		1	7	3	11
TU Eindhoven		1	2	1	4
TU Twente		0	0	0	0
RU Groningen		2[a]			2
Open Univ.		1	1		2
RU Leiden	2[a]	4			6
RU Utrecht	1			1	2
Erasmus Univ.	2[a]				2
RU Limburg	3				3
KU Nijmegen	4	2			6
Totals	17	12	10	7	46
Qualifying Respondents	9	3	2	1	15[b]

a. Emeritus
b. Of the total, 2 respondents were emeritus and 4, although tenured, were employed part-time.

Because the responses to the questionnaire were anonymous, it was not possible to determine the university affiliation of the individual respondents.

chemistry (all). Of the forty-six questionnaires distributed, eighteen were returned (three of which were disqualified).[2] Table 9.1 illustrates the distribution of the population[3] under consideration, together with the distribution of the respondents.

2. In order to keep the sample as representative as possible, three of the returned questionnaires were disqualified because the individuals, although trained in exact science fields, were currently engaged in (broadly defined) social science research (e.g. physical anthropology).
3. Abbreviations of universities:
UvA = University of Amsterdam
VU = Free University
TU Delft = Technical University at Delft
TU Eindhoven = Technical University at Eindhoven
TU Twente = Technical University at Twente
RU Groningen = National University at Groningen
Open Univ. = Open University
RU Leiden = National University at Leiden
RU Utrecht = National University at Utrecht
Erasmus Univ. = Erasmus University at Rotterdam
RU Limburg = National University in Limburg
KU Nijmegen = Catholic University at Nijmegen

Discussion

Current Educational Factors

Educational careers of boys differ from that of girls in the Netherlands. This difference is most clearly evident when students are confronted with constructing their study programme in preparation for future career choices. At this level girls often choose areas in the humanities, while boys more often opt for the exact sciences. (Van Oost, 1986)

Up to 1977 there were fewer girls than boys in VWO (Academic Preparatory Education). Although the numbers were more equal after 1977, fewer girls completed the VWO; and then, as now, fewer girls than boys opted for the exact sciences study package. When choosing their study orientation at secondary school level, girls take longer and are more serious about their decisions. Dekkers (1985) attributes this to their confrontation with the fact that they might ultimately have to choose between family and a career – particularly when their interests lie in the direction of the exact and technological sciences. This quandary does not exist for boys.

During the late 1970s and early 1980s, a number of studies were initiated on researching gender bias in the schools (e.g. Lubbers and Menting, 1985; Kluvers and Goedhart, 1986). The results of these studies indicated that girls were under-represented in the exact and technical sciences. Similarly, teachers often underestimated the knowledge of girls; reported their performance level as often being below the norm; and were more inclined to interact with male students in exact science courses than with female students. Not surprisingly, such studies report that girls have low self-esteem in science fields. This is further exacerbated by the paucity of female mathematics and science teachers; the freedom to choose a career-oriented study curriculum at secondary school level; and the perception girls have, and retain, of the exact sciences as 'male' domain professions. (Goedhart, 1986)

In a 1986 study, Goedhart evaluated three studies designed to research the factors determining the career orientation and curriculum chosen by girls. She found that male and female students tend to be equally influenced in choosing a science career by those teachers (of either sex) who actively stimulate and encourage them to 'choose exact'. Nevertheless, a far more important factor in influencing the career choices of both boys and girls may be the Dutch alternative curriculum programme. In the Netherlands, as in many developed countries, students are able to choose independently their field of study. In such a system teachers and advisers play an important, but primarily influential role. (Van der Hoek, 1987) This is in sharp contrast to, for example, the Turkish mandated study system; and to the effect of such a system on the number of women entering science and technology fields.[4]

Nevertheless, the Dutch VHTO (Women in Higher Technical Education) continues to support the position that 'socialisation and information' is the best strategy for encouraging and stimulating girls' interest in science and technology. Unfortunately, this strategy does not appear either to stimulate girls to enter

4. In 1983, a total of 1,406 girls were studying engineering (excluding architecture). This comprised 7 per cent of the total number of students enrolled in technology fields. In this respect, The Netherlands compares favourably with other highly industrialised countries, e.g. in England, West Germany, and Japan women comprise, respectively, 4 per cent, 5 per cent, and 2.5 per cent of the total students in engineering fields (figures from 1981/2). Compared to a group of ten developing countries, however, these percentages are low. In the developing countries referenced by Van der Hoek (1987), the percentage of women engineering students is approximately 12 per cent, with women comprising 15 per cent of the total number of students in technology studies (these figures exclude architecture and landscape architecture).

It is interesting to compare, for example, the effects of the Dutch alternative curriculum programme with the Turkish mandated study system. In 1976, in Turkey, there were approximately 7,000 women studying in technology fields, and by 1983 there were 10,210 (the percentage of men enrolled for this same period decreased slightly). The most popular fields for women in Turkey are medicine and law, with growing interest in electrical engineering and computer science. Although economic factors are contributory, the educational system is the real underlying force behind these figures. (Van der Hoek, 1987) Every year university entrance exams are held nationally; the resulting scores (40 per cent of the exam comprises mathematics) will determine both who goes to university, and what they will study.

This is the opposite of what happens in the Netherlands and other developed countries, where students can choose their own field of study independently. (Van der Hoek, 1987) It must be pointed out, however, that in Turkey the exigent and increasing need for specialised manpower, in tandem with a powerful social elitism, underpins this relatively recent 'plethora' of professionally employed women. It is the women from the elite class, rather than men from the lower classes, who are stimulated to enter university and compete for high-ranking professional positions. This effectively closes the ranks of the elite class, keeping both economic prowess and social status a 'family affair', while simultaneously stimulating women to cultivate an interest in the fluctuations of the labour market. (Van der Hoek, 1987)

these fields, or actively to cultivate their interest in the fluctuations of the labour market. Moreover, if the questionnaire results can be taken as representative, most young women who opt for an exact or technical study package do not need to be convinced; the majority of respondents in this study experienced a 'natural' affinity for their chosen field by secondary school level. However, many respondents also pointed out that it was their persistence – often in the face of considerable resistance – which ultimately afforded them a career in their chosen field.[5] This attitude, together with a propensity for permanent career plans, may have been facilitated by the parental (more often maternal) stimulation and support received by most of the respondents. Interestingly, however, with few exceptions the majority of the respondents' mothers had not had careers. This may, in part, explain both why most of the respondents felt that a career in research and a family is a difficult combination, and why the majority were unmarried or divorced with no children. An additional factor may be that most of the respondents worked irregular hours with an average working week comprising between forty-five and fifty-five hours.

Historical Deterrents to the Education of Women

The obstacles women encounter in entering science and technology fields are, of course, exacerbated by the general problems women have always had in gaining access to all forms of male dominated education and public sector professional employment. In Europe this was, to a great extent, related to the exclusion of women from the entire educational process. Moreover, the historical development of university institutions and the limited access to membership therein (on all levels) has, until very recently, constituted an overt, and usually insurmountable, obstacle for women.

Not until the end of the nineteenth century was a woman openly enrolled as a full-time student at a Dutch university; in 1879 A. Jacobs obtained a Ph.D. in medicine. Although others followed, the research endeavours of the majority of Dutch

---

5. This may, however, be relative to the time frame within which most of the respondents were completing their higher education and entering the academic labour market (the average age of the respondents was forty-five years).

women academics in exact and technical science fields in the early twentieth century fell on infertile ground. More often than not they were passed over for high level academic positions. Clearly, the adage of a premium on excellence was extended only to the male species in the halls of academe (see for example Van Loosbroek *et al.*, 1988).

Although times have changed considerably with respect to entry into the university and the availability of employment and promotional opportunities for women in the exact and technical sciences, the – albeit increasingly covert – social stigma associated with women opting for professional careers continues to be evident in the high 'social costs' women must pay for their choice. This is evident not only in the – still negligible – number of women in exact and technical fields, but also in the high percentage of women in these fields who neither cohabit/marry nor raise families.

Women in the Exact and Technical Sciences

The only technical university in the Netherlands prior to 1956 was at Delft.[6] Between 1947 and the establishment of the third technical university in the Netherlands in 1964, the number of women enrolled at Delft remained roughly constant. This has been attributed to such factors as a general ineptness and consequent lack of interest among girls for science and technology professions, and a social environment which dictated neither the necessity, nor the desirability of higher education for girls. Preparation for marriage and motherhood had priority over preparation for gainful professional employment.

The fields of specialisation chosen by women also remained constant during this period. Very few women studied physics, civil engineering, electronics or mechanical engineering. (Gillissen and Lissenberg, 1985)[7] Although the number of women enrolled at the technical university at Delft did begin to increase

---

6. A second technical university was established in Eindhoven in 1956, and a third in Twente in 1964.

7. Those science and technology fields most preferred by women in the 1950s (and also in the 1970s) were chemistry and architecture; and in the early 1950s approximately 190 women completed their studies in chemistry and related technologies. (Gillissen and Lissenberg, 1985) It should be noted, however, that women in these areas gravitated to the 'softer' sectors of their chosen field.

Table 9.2: Total (full-time, tenured and untenured) professorial and lecturer positions at the three major technical universities in the Netherlands.

University	Professorships Total	Women	Lecturers Total	Women
TU Delft	208	3	207	2
TU Eindhoven	118	1	92	2
TU Twente	88	0	76	1[a]

Source: Feiten & Cijfers, Tweede voorpublicatie, Ministry for Science and Education, 31 May 1989.

a. In the 'Department of Education'

after 1964, by 1986 there were still only about 1,000 women studying in Delft (roughly 10 per cent of the total student population at that university). Unfortunately, the numerous initiatives taken by this, and other Dutch technical universities in recent years have not substantially increased the total percentage of women enrolled at such institutions.

Academic employment figures for women in exact and technical fields in the Netherlands are also low. For example, in 1980, 97.3 per cent of the research scientists employed at the TU Delft were still men. Of the total female scientific staff (2.7 per cent), the majority were in the field of architecture. Gillissen and Lissenberg (1985) mention that in the 1950s there was only one female professor at Delft (in architecture).[8] Currently, there are three female professors at Delft – one in architecture, one in industrial design, and one in chemical technology. (Gillissen and Lissenberg, 1985; De Jong, in 't Hart, 1986; Hicks and Noordenbos, 1990)

As noted above, women continue to be under-represented in professorial and lecturer positions in the exact and technical sciences. Table 9.2 indicates the total number of women currently holding professorial and lecturer positions at the three major Dutch technical universities.

---

8. De Jong, however, lists A.E. Korvezee as the first professor in Delft (in 1947 in theoretical chemistry). It may be of note that Korvezee had a 'buitengewoon' professorship, i.e. externally funded. De Jong adds that her appointment may have been in conjunction with the general shortage of post-war academicians. (De Jong, in 't Hart, 1986)

The Role of Women's Organisations and Emancipation Programmes

Women's Organisations An important factor contributing to the minimal participation of women in the university generally, and exact and technical fields in particular, may be that women lack a formal network support system. (Outshoorn, in Leijenaar, 1987) Indeed, in the Netherlands, only during the last two decades have women increasingly organised themselves into groups.[9] Women's student organisations have fared no better. One of the first – and only – formal women's student organisations, the DVSV (the Delft women students organisation) was founded in 1904, with thirteen members. Up to 1970 almost all women enrolled at TH Delft were members of this organisation; but in 1976 the DVSV fused with the DSC (The Delft Student Corps) due to lack of membership and funding. (Gillissen and Lissenberg, 1985) Apart from the DVSV, the Netherlands had no equivalent of such professional organisations as the British Women's Engineering Society (WES), which continues to remain active. Nor can the Netherlands boast a tradition of women's colleges.

Although independently active, the women's organisations currently in existence have been ineffectual in increasing the number of women opting to study the exact and technical sciences. Up to now they have been primarily involved with women either currently studying at higher education institutions, or already professionally active in these fields. Only recently has an attempt been made to motivate young women at the level of secondary education to go on to study in exact and technical fields. Unfortunately, such organisations can only be effective when – and if – they have significant and representative membership. (de Jong, in 't Hart, 1986)

While such institutions as women's colleges and organisations may not directly influence the number of women in exact and technical science fields, they can provide the social skills necessary to rise to the top of the scientific community (such

---

9. Examples include Vrouwen en Bouwen, Bêta-Vrouwen, Vrouwen in Technische Beroepen, HTS-vrouwen, Meiden op de TH, en Vrouwen en Informatika (Women and Building, Bêta-women, Women in Technical Professions, Women at Technical Colleges, Girls at the Technical University, Women in Communication Theory). (For de Jong, see 't Hart, 1986)

skills include for example leadership, management and fundraising experience, research collaboration, professional assertiveness and initiative). It is illustrative that while the majority of respondents in this study felt that women must be more qualified than their male counterparts to compete for jobs, they also agreed that women are generally less concerned with wages and promotion than are their male colleagues.

The problem of women in science and technology clearly involves more than merely increasing their numbers (and the number of girls studying the exact sciences at school). As the sample in this study indicates, even when women do enter these fields they only rarely reach the top positions. The hypotheses posited for why this might be so include for example: the paucity of women in these fields; the small numbers of girls opting to study these fields; the rigours of juggling marriage and family with career, together with family-related career interruptions; the lack of labour market sophistication and late entry into same; no network development, and women's lack of assertiveness and initiative (see for example 't Hart, 1986; Beekes and van Doorne-Huiskes, 1988; Hicks and Noordenbos, 1990).

This study certainly indicates the continuing paucity of women in these fields, with many of the respondents being the only woman in a top position in their department; and in many cases being only one of a negligible number of women employed by the department. Similarly (as indicated above) there does not seem to be a substantial increase in the number of girls opting for exact and technical science fields at secondary school level. Ironically, as most of the women in this study are either unmarried or divorced without children, they are implausible role models for the potential combination of career and family. Indeed, the majority were of the opinion that a research position and a family is a difficult to impossible combination.

The claims of low labour market sophistication and late entry are not, however, substantiated by this sample. A majority of respondents reported having experienced positive employment and promotion opportunities throughout their career. Nevertheless, although more than half of the respondents acquired their latest appointment by invitation, this may be equally correlated to professional qualifications and emancipation policy pressure.

Unfortunately, the relatively recent Dutch tradition of the

alternative of part-time employment, coupled with the fact that women far exceed men in holding part-time and/or temporary positions (indeed often seem to opt for them) must ultimately take its toll on the potential for women to compete on the open labour market. While some part-time employment is research project-related, the majority of tenure track part-time positions constitute teaching appointments. Part-time and temporary employment all too often precludes taking part in project-related funding activities; an important factor in becoming part and parcel of the information/opportunity network. Moreover, the combination of a paucity of women actively engaged in exact and technical science research fields and the increasing number of women opting for – or constrained to accept – part-time and/or temporary positions constitutes a formidable obstacle to the formation and development of an independent, albeit consociate network structure competent to interface with the existing institutional and information structure. A more recent factor affecting women's opportunities at university level may prove to be the institution of the, minimally paid, Ph.D.-level teaching and research assistantships. Although it is too early to determine whether more women than men will hold such positions, it is the case that more men than women in exact and technical science fields are opting for industrially related employment, potentially leaving such assistantships and university positions open to women candidates. The ultimate employment opportunities and obstacles generated by this situation will be a matter for future evaluation. The actual degree to which women in exact and technical science fields do – or do not – constitute assertive initiators will also be a matter for future research.

Emancipation Programmes Initiatives have been taken at all levels actively to involve more women in the exact and technical sciences; government has sponsored bonuses to companies hiring women for technical positions, the EEC has passed legislation to force companies and universities to hire more women in these areas, and Dutch technical universities have activated programmes designed to attract more female students.

Additionally, during the last ten years the Dutch Ministry of Education and Sciences has adopted an extensive, and widely supported, emancipation policy. Although this policy has shown results at the lower levels of secondary education, it has not significantly affected the number of girls choosing to study the exact technical sciences at the higher secondary school level. (Van Oost, 1986) This has clear implications for the number of girls who will go on to higher levels of education in the exact and technical sciences, and of course the number of women who will be professionally active in these areas.

Conclusions

If we can assume that women's participation in exact and technical science fields at the academic level is indeed correlated both to the number of women already employed in these fields and the status/rank positions they occupy, then we can only conclude that the participation of women and their potential to engage in decision-making policy in the Netherlands is minimal, at best. An overview of the relatively recent history of large numbers of women entering the labour market in the Netherlands indicates that this late entry may be an important fact contributing to the under-representation of women at all levels in academe. Nevertheless, this is by no means the only impediment to increasing the numbers of women in decision-making positions in exact and technical science fields.

Although deterrents, both to the admission of women to exact and technical science programmes at the university level and to their activity on the labour market, are no longer overtly manifest, socio-cultural deterrents continue covertly to inhibit women's participation in these fields. This is clearly illustrated by the limited financial support for those strategies posited by the Emancipation Commission to increase the number and position of women on the labour market.[10]

Because the household duties women perform do not constitute paid labour, such duties are not normally viewed as work. When women with families also become gainfully/profession-

---

10. These strategies include, for example, increasing child care facilities, supplementary education for women, job and salary redistribution through a decreased working week for both sexes, and adequately paid part-time employment at all levels and for both sexes ('adequately paid' implies self-sufficiency). Such changes are especially relevant for younger women already employed in science and technology fields, where a sixty-hour work week is rapidly becoming the standard.

ally employed, or the converse, they have both the rigours of professionalism and the household to contend with. Such women, more often than not, find themselves confronted with choosing between home and profession. Verplanke clearly illustrates the unfortunate results of this problem of 'choice'. (1986) She is a faculty member in the Biology department at the University of Groningen; a department which has a large number of women on the faculty. The majority of these women are neither married, nor cohabiting, nor do they have children. She points out that the experiment-related fields in which these women are engaged are not 'nine to five jobs'; a fact which generally precludes having two jobs (in this case that of scientist and wife/mother). Her findings are substantiated by the results of the questionnaire utilised by this study. However, the fact that the majority of women in the sample were either unmarried or without children does not constitute a platform for arguing against increasing opportunities for actual or potential working mothers; nor does it diminish the relevance of reform programmes primarily aimed at 'household' problem-solving. On the contrary, many girls at secondary school level do not opt for science careers precisely because of the serious social constraints this will place on their lives and, by extension, their perceived 'social' choices. We might even conclude that such constraints continue to appear paramount to girls of this age group precisely because they are neither applicable to nor experienced by boys.

Compounding the problems of academically professional women has been their inability compositely to interface with the established, rapidly developing 'knowledge' structure and to engender and nurture the professional collaborative and network contacts necessary to compete at both the intra- and international professional level. By extension, entrance into the existing 'knowledge' structure is requisite to developing the basic professional (e.g., access to funding and professional – hierarchical – collaboration) and personal (e.g., developing contacts, management and decision-making prowess) skills necessary to aspire to and compete for higher echelon decision policy-making positions.

Attempts to mitigate these circumstances in the Netherlands have primarily taken the form of the active development, during the last two decades, of women's professional organisations and university-based women's studies programmes. The latter, especially, were ostensibly established to research the continuing under-representation of women in academe, proffer potential solutions, and actively support the process of instituting emancipation policy and related strategies.

By 1986, however, Bêta-women's studies (exact and technical science fields), in tandem with their Anglo-Saxon counterparts, were also expressing concern about the nature of science and technology research; questioning the inherent structure and research paradigm orientation of science and technology fields. Unfortunately, the alternative approaches proffered to date are too often scientifically untenable or marginal at best, and are insufficiently operationalised (or cannot be operationalised). This has had the unfortunate consequence of further alienating those women engaged in exact and technical (and other academic) careers both from women's studies and the feminist movement.

Recently initiated research projects at various Dutch universities hope to provide insight into the success – or futility – of women's professional organisations and university-sponsored women's studies programmes, and of the potential contribution to scientific knowledge of their 'alternative' approaches to research, and their capacity to effect change (current and potential) in the social and professional position of women on all levels. Ideally, the more successful women's organisations and studies programmes are in initiating and orchestrating emancipation programmes and related strategies, the greater will be their influence in expanding career opportunities for women generally. Moreover, the expansion of such opportunities will undoubtedly, albeit in the long run, influence the curriculum and career choices of girls at secondary school level, i.e. their perception of the feasibility of having a viable career in an exact or technical field. Equally importantly, if such opportunities are buttressed by the implementation of those strategies suggested by the Emancipation Commission, then we may yet see the slow, but thorough, erosion of the norm that women have sole responsibility for family and household maintenance. Consequently, girls at secondary school level would no longer be confronted with the alienating choice between home and profession.

A final, but essential problem continues to be the inequality between men and women; both of whom are culturally relegated to role patterns neither defines. Both are socialised into their aspirations, expectations, and the perception both of their 'career' alternatives and their role and behaviour pattern therein. This is a reflection of the traditional public/private sphere relegation of the activities of men and women. Consequently, as long as women perceive themselves as constituting a low-status alienated minority at all levels of the work force, it is the men who will continue to set the standards and control hiring practices, thereby inadvertently reinforcing the status quo.

References

BEEKES, A., 'Meer Vrouwen, Meer Problemen' (More Women, More Problems), in W. van Rossum, E.K. Hicks, J.C.M. van Eijndhoven, (eds.), *Onderzoek naar Wetenschap, Technologie en Samenleving*, (Science, Technology and Society). Amsterdam: SISWO publication, 1987, pp. 11–19.

BEEKES, A. AND A. VAN DOORNE-HUISKES, 'Vrouwen aan de universiteit: Een toets van Kanter's tokentheorie' (Women at the university: a test of Kanter's token theory), in *Mens en Maatschappij*, vol. 63, no. 2, 1988, pp. 175–87.

Bêta-Vrouwen Congres, Bêta-women's conference report, 22–23 April, 1988, Leidse Stag-drukkerij, Leiden.

BOUMAN, I., 'Scenario Arbeid' (Scenario, Work), in *Kwartaal Nieuws, Emancipatieraad*, 22, Sept. 1987, pp. 4–6.

DEEG, D., 'Zijn Betavrouwenstudies het hobbyism ontgroeid?' (Have Bêta women's studies outgrown the 'hobby' stage?), in *Wetenschap en Samenleving*, 8, Nov. 1987, pp. 17–22.

DEKKERS, H., 'Keuzeprocessen bij meisjes: Invloed van met name keuzebegeleiding door dekanen en vakkenpakketten en school-type keuze van meisjes' (Educational choices for girls: influence of guidance counselling, study specialisation and girls' choice of schools). Nijmegen: ITS, 1985.

———, AND M. SMEETS, 'Sekse-ongelijkheid op school (I & II)' (Gender inequality in schools (Parts I and II). Nijmegen: ITS, 1982.

Derde nota Onderwijsemancipatie: (Third NOTA for educational emancipation). Government publication, Den Haag, 1983.

DIEDEREN, J., 'De keuze van een beroep: van jaar tot jaar, 3d fase' (The choice of a profession: from year to year, 3d phase). Nijmegen: ITS, 1983.

VAN DOORNE-HUISKES, J., 'Vrouwen en beroepsparticipatie: een onderzoek onder gehuwde vrouwelijke academici.' (Women's professional participation: research on married female academics). Utrecht, 1979.

——, 'Wage Differences Between Women and Men in Academia', in *The Netherlands' Journal of Sociology*, vol. 24, no 2, October 1988, pp. 146–58.

DIJK-DEN BANDT, M.L., 'Akademisch gevormde vrouwen en hun mogelijkheden' (Opportunities for women with higher education). *Mens en Onderneming*, 26, May 1972, pp. 181–93.

ENZING, C., 'Vrouwen in de natuurwetenschappen' (Women in the exact sciences), *Tijdschrift voor Vrouwenstudies*, vol. 1, no. 1, 1980, pp. 94–9.

EVERTS, S. AND E. VAN OOST, 'Vrouwelijke studenten aan de TH Twente. Een verkenned onderzoek naar de studieloopbaan van vrouwen in technische studierichtingen' (Women students at the Technical University at Twente: A study of the educational careers of women in technical fields). Enschede: TU Press, 1985.

GILLISSEN, A. AND A. LISSENBERG, 'Eindverslag van de stage Meisjesstudenten in Delft' (Final report of the practicum on female students in Delft). Amsterdam, unpublished paper, 1983.

——, 'Vrouwen in een mannenbolwerk' (Women in the male bastion: experiences of women students in the 1950s and 1970s at the Technical University at Delft). Sociological Institute, University of Amsterdam, 1985.

GOEDHART, V., 'Factoren in enige onderzoeken naar exakte-vakkenkeuze van meisjes' (Factors in selected studies on exact science field choices made by girls). Ment-Report 86–05, Didaktiek Natuurkunde, TH Eindhoven, 1986.

'T HART, J. et al., (eds), 'Een barst in het bolwerk' (A breach in the bulwark: women, exact sciences and technology) Selected papers. Amsterdam: SUA, 1986.

HICKS, E.K., 'A Commentary on Women's Participation in the Exact and Technical Sciences in The Netherlands', in *WO NieuwsNet*, no. 4, Dec. 1988, pp. 22–8.

——, 'You can't be an engineer and still go to church', in *WO Nieuws-Net*, 5, Spring 1989, pp. 11–23.

——, AND G. NOORDENBOS, 'Is Alma Mater Vrouwvriendelijk?' (Does Alma Mater Discriminate?). Assen: Van Gorcum b.v., 1990.

VAN DER HOEK, J., 'Derde Wereld loopt voorop' (The Third World is in the lead), in *Intermediair*, no. 23, 12, 20 March 1987, pp. 39–41.

JANSEN, L., et al., 'Literatuuroverzicht van de positie van vrouwen en het emancipatiebeleid in 1986.' (Bibliography: the position of women and emancipation policy in 1986). The Hague: Ministry of Social Affairs and Labour, 1987.

JUNGBLUTH, P., 'Docenten over onderwijs aan meisjes: positieve diskriminatie met een dubbele bodem' (Teacher's views on educating females: positive discrimination and a double standard). Dissertation, Institute for applied sociology, Nijmegen, 1982.

KLUVERS, I. AND V. GOEDHART, 'Meisjes apart voor exacte vakken? Hoe apart zijn meisjes?' (Separate exact science instruction for girls? How separate are the girls?). The Hague: Ministry of Education and Science Publication, 1986.

KOENEN, G., 'Vrouwen in mannenbanen: Een nieuwe arbeidsdeling?' (Women in male professions: A new division of labour?) Sociological Institute, University of Amsterdam, 1985, 2nd edn.

LEIJENAAR, et al., (eds), The Gender of Power. Proceedings of a symposium; Vakgroep Vrouwenstudies, FSW, Leiden, 1987.

LIEBRAND, J., 'Emancipatie: een ondergeschoven kindje. Facetbeleid geen gelukkige keuze' (Emancipation: a spurious child. The Facet Policy proves an unsuccessful choice), in Opzij, 16, 2 Feb. 1988, pp. 30–31d.

VAN DE LINDEN, P. AND J. VORENKAMP, 'Onderzoek naar invloed van sekse-verschillen op beroepskeuze' (Research into the influence of gender on choice of profession), in Didaktie, March 1984, p. 29.

VAN LOOSBROEK, et al., (eds), 'Geleerde Vrouwen'. (Learned Women) Negende Jaarboek voor Vrouwengeschiedenis. SUN, 1988.

LUBBERS, J. AND C. MENTING, 'Kortsluiting in de interaktie: een empiries onderzoek naar seksspecifieke selektiemechanismen in de leerkracht leerling interaktie bij natuurkundegroepswerk' (Interactional short-cuts: empirical research into sex specific selection mechanisms in teacher-student interaction in natural science discussion groups). Thesis, Institute for General Linguistics, University of Amsterdam, 1985.

——, 'Girls and Science Education: Selection in Classroom Interaction', in D. Brower and D. de Haan (eds), Women's Language, Socialization and Self-image. Dordrecht: Foris, 1987, pp. 114–26.

MARTENS, A., F.C. VALKENBURG, AND A.M.C. VISSERS, 'Theorie van de dubbele arbeidsmarkt. Samenvatting' (Theory of the double labour market: a summary). National Programme for Labour Market Research, Publication No. 5. The Hague: Nl. Govt. Publishing, 1979.

Ministry of Science and Education, 'Beleidsnota Voorbereidend Hoger Onderwijs' (Policy report on preparatory higher education). The Hague, 1985.

——, Feiten & Cijfers (Facts and Figures) Tweede voorpublikatie, 31 May 1989.

NIEUWENBURG, C.K.F. AND J.J. SIEGERS, 'Naar een gee mancipeerde arbeidsverdeling' (I en II) (Toward an emancipated labour distribution) (I and II). Economisch Statistische Berichten 66, no 3311 and 3312, 1981.

NOORDENBOS, G., 'Institutionele uitsluiting van vrouwen uit wetenschappen' (Institutional exclusion of women from the sciences). Dept. of Phil. of Science Report, RU Groningen, 1982.
——, 'Vrouwen in het Wetenschapsbedrijf' (Women in the science business), in W. van Rossum, E.K. Hicks and J.C.M. van Eijndhoven, (eds), *Onderzoek naar Wetenschap, Technologie en Samenleving* (Science, Technology and Society). Amsterdam: SISWO, 1987, pp. 20–5.
NVON, Proceedings of a 'discussion day' on women in the natural sciences, April. Contact group Emancipation in State Education, Ministry for Science and Education, The Hague, 1985.
VAN OOST, E., 'Etude in B-mineur. Een literatuurstudie naar verschillen tussen meisjes en jongens bij keuze van exacte vakken in het voortgezet onderwijs' (Etude in B-minor: a literature study of the differences between girls and boys in choosing the exact sciences at the secondary school level). TH Twente, 1986.
DE RAAF, I. AND M. VAN VONDEREN, 'De Technische Hogescool als studiekeuzemogelijkheid voor meisjes' (The Technical University as alternative for girls). Eindhoven, 1981.
RANG, B., 'Geleerde vrouwen van alle Eeuwen en de Volckeren, zelfs oock bij de barbarische Scythen: De catalogi van geleerde vrouwen in de zeventiende en achttiende eeuw.' (Learned women from all centuries and peoples, even among the barbaric Scythians: The catalogue of learned women in the 17th and 18th centuries), in Van Loos, *et al.*, (eds), *Geleerde Vrouwen, Negende Jaarboek voor Vrouwengeschiedenis*, SUN, 1988, pp. 36–44.
'Scenario Arbeid' (Emancipatieraad), Emancipation Advisory Commission Report. Policy suggestions with respect to girls and young women on the labour market. The Hague, 1987.
SAHARSO, S. AND J. WESTERBEEK, 'Schets van een beleid voor emancipatie in onderwijs en wetenschappelijk onderzoek' (Outline of an emancipation policy for education and scientific research), 1979.
Ministry of Education and Science: 'De precaire balans. Een reconstructie van de ervaringen van vrouwelijke studenten met studeren, achterraken en uitvallen' (A precarious balance. A reconstruction of the experience of female students with their study, falling behind and dropping out). Amsterdam, 1983.
SCHOENMAKER, N., A.M. DE JONG-VAN DER POEL AND R.W. HOMMES, 'The Position of Women on the Labour Market: An exploration of research areas'. Amsterdam: SISWO, 1978.
Steungroep Vrouwen in Technische Beroepen (Support group for women in technical professions). Report, Amsterdam, 1981.
Technical University Delft, Nota emancipatiebeleid (Emancipation policy report). Delft, 1981.

Technical University Eindhoven, Statistieken aantallen studenten 1975–1981 (Statistics on the number of students enrolled between 1975–1981) Eindhoven, 1982.

VERPLANKE, J.J., 'Vrouw of natuurwetenschapper?' (Woman or natural scientist?), *Intermediair*, vol. 11, no 14, 4 April 1986, pp. 55–61.

VERWAIJEN, I., 'Study uitval en vertraging van vrouwen in het hoger onderwijs' (Women dropping out and procrastinating in higher education). The Hague, 1985.

VHTO: (Foundation for Research on Women and Technical Education) Meer vrouwen op de HTS (More women in Technical Colleges). Proceedings from a symposium held in June, 1986. Hengelo, 1986.

VAN VONDEREN M. AND I. DE RAAFF, De Technische Hogeschool als studiekeuzemogelijkheid voor meisjes (The Technical University as higher education choice for girls). THE, Vakgroep Gedragswetenschappen, 1981.

Vrouwenstudies in de Natuurwetenschappen (Women's studies in the exact sciences). Full edition of *Wetenschap en Samenleving*, no. 4, April 1982.

'Wetenschapsbudget, Tweede Kamer der Staten-Generaal, Vergaderjaar 1988–1989, Meerjarenplan voor wetenschapsbeoefening'. (National science budget covering the period 1985–1989). Government publication, The Hague, 1989.

'Wetenschaps- en technologie -indicatoren' (Science and technology indicators). The Hague: RAWB 1988.

WISSE, A., 'De vrouw en de universitaire studie; (Women and university study), in *Leven en werken*, jrg. 2, July 1988, pp. 1–15.

10

Equal Opportunities for Women?
Women in Science in Hungary

Agnes Haraszthy

General Position of Women in Science in Hungary

The process of industrialisation in the late nineteenth century was accompanied by the appearance of women in increasing numbers in the intellectual professions. By the 1930s the number of women in the intellectual fields became more and more significant in Hungary, too.

The country's socio-political change after 1945 offered greater advancement opportunities for women, although – from the point of view of women – no real breakthrough was taking place. Women are especially disadvantaged in the creative intellectual professions, although the proportion of women among graduates from higher education institutions is relatively high. Trends in the proportion of women graduates in different fields are shown in Table 10.1.

Despite the existence of a relatively great reserve of women with higher education (40 per cent of the graduates in 1980), certain positions, requiring higher qualification and total commitment are still unattainable for women.

The pyramidal structure of women's occupational distribution is well illustrated by the education sector. In 1984, women constituted 99 per cent of the kindergarten personnel, yet only 20 per cent of the primary school teachers and 33 per cent of university teachers. The ratio of women is especially low at the higher levels of state administration. Women's limited advancement opportunities are not due to their lack of abilities or skills. This situation is the result of the traditional gender-division of labour, of the

Table 10.1: The share of women graduates from higher educational institutions (at the beginning of the year)

Field of studies	% women			
	1949	1960	1970	1980
Technical sciences and engineering	1.2	7.3	11.7	18.9
Agricultural sciences	4.2	9.8	13.5	17.1
Medical sciences	17.2	26.6	37.6	47.7
Social sciences	27.2	29.1	35.6	43.2
All graduates	16.6	23.0	31.2	40.0

Source: Adatok a nök hejzetéröl (Data on the Position of Women). Budapest, Central Office of Statistics, 1982.

operation of the labour market, as well as women's poorer chances to promote their individual interests.

It is a rather strange situation that no less than two generations have grown up in Hungary on the basis of the ideology of emancipation, yet the traditional image of men and women has remained practically unchanged. This is evident from a number of studies in several fields that range from the analysis of textbooks to the man/woman image as presented in the mass media . (Sas, 1984) For example, it is women who in television commercials usually deliver the message about what should be purchased or what the best product is. It is women who tell bedside stories for children, while there are hardly any women among the commentators on foreign policy. Similarly, commentators or experts on scientific topics – except perhaps in the life sciences – are mostly males. Although the ideology of the equal rights of women has spread considerably, the models derived from everyday life fall short of it. Thus neither of the two sexes benefits from this.

Starting from the 1950s along with the development of the R & D system, there was a marked increase in women's involvement in science. By the mid-1960s women accounted for one-fifth of the total R & D personnel. Although their proportion has been constantly increasing, it still lags far behind the share of women in other intellectual fields.

Currently, more than one-quarter, or 28.3 per cent of all R & D personnel are women. Their distribution by scientific fields differs widely: in the social sciences their share is 40 per cent (within this: in psychology 55 per cent, in pedagogy 43 per cent); in medicine 34 per cent, while in engineering sciences not more than 23 per cent. The differences in the proportion of women between the individual branches of science may be explained by several factors. Rapidly expanding disciplines are most likely to attract some marginalised groups (thus women, too). On the other hand, different fields may be characterised by varying degrees of conflict between the demands of research activities and the traditional female roles. The fewer conflicts there are in any given field, the higher women's participation.

Women scientists also fall into two distinct age groups. Since women were embarking on scientific careers in gradually increasing numbers in the late 1960s, there seems to be a relative predominance of those between forty to fifty years of age among the researchers. On the other hand, there is a group of 'latecomers', who are younger than the average for the scientific community.

Women in the Hungarian Academy of Sciences

In 1976 – and not since then – the Hungarian Academy of Sciences put the situation of women scientists on the agenda, in order to examine the possibilities of furthering the participation of women in scientific life, and to see what measures were to be taken to reduce the obstacles presented by objective circumstances against the professional progress of women. The Presidium of the Academy emphasised that women scientists claim no special favours or benefits, but equal chances with their male colleagues to bring their creative abilities into full play, while taking into account their objective biological characteristics, too. It was felt important that the research institutions should provide equal chances during the first ten years following graduation, since this is a very formative period for scientists.

Nevertheless, the more prestigious research institutes of the Hungarian Academy of Sciences employ a comparatively smaller number of women on their research staff than other research institutes in corresponding fields. Women accounted for one-quarter, or 676 of the total of 2,833 scientists employed by these institutes in 1988. (Hungarian Academy of Sciences, 1988) One-

Table 10.2: Gender differences in scientific productivity among scientists in natural and technical sciences, 1980–5

	\multicolumn{6}{c|}{Average number of products for 1980–85 period}					
	\multicolumn{3}{c|}{Produced alone}	\multicolumn{3}{c|}{With co-author}				
	Men	Women	Total	Men	Women	Total
Books	7.9	7.0	13.0	4.9	11.0	—
Patents	3.9	0.6	2.9	25.8	8.6	21.2
Innovations	4.1	0.6	3.1	10.6	1.4	8.1
Reports	30.2	25.4	28.7	52.7	40.9	49.3
Articles	36.4	23.1	32.7	51.4	41.8	48.7

Source: Tamàs, 1984.

quarter of these have a scientific degree.[1] The number of women with only university degrees is very large. The main reason for this is the lack of conditions necessary to link the double responsibility or role of women – work and family. In Hungary this poses a particularly serious problem in the case of occupations with requirements higher than the average.

Women's scientific achievements are more moderate than those of their male colleagues. This is reflected in the gender differences in the age of obtaining scientific degrees. The age of women at the time of obtaining the 'candidate' of science degree ranges from forty-one to forty-seven years. Of the twenty-three women with 'doctor of sciences' degrees only two are in their forties, the rest are over fifty. Men are in general three to four years younger. Although in principle women have equal chances with their male colleagues for academic advancement, their number does not increase within individual fields of science according to their proportion. Obtaining the candidate or doctoral degree requires much more energy from the woman than from the man. Accordingly, male scientists are able to defend their theses earlier than women, who are prevented by their other, family obligations.

---

1. The scientific degrees are the lower level 'Candidatus Scientiarum' (C.Sc.) and the higher level 'Doctor Scientiarum' (D.Sc.) The C.Sc. degree requires that the applicant passes certain prescribed examinations before handing in his/her dissertation which has to be defended in a public debate before a panel of appointed scholars. To obtain a D.Sc. degree, a scholar must submit new findings based on original research and defend the thesis in a public debate.

Differences between the sexes are most conspicuous in the area where the results of scientific activities appear in measurable products. In such areas, the advantage of the men is clearly marked and the success indicators of the two sexes are highly different.

On the other hand, gender differences are equally striking with respect to organisational promotion opportunities within the research field. Not one of the 165 eligible women with scientific degrees can be found among the heads or deputy heads of any of the research institutes of the Hungarian Academy of Sciences. Even at the lowest level of the organisational hierarchy, not more than thirteen women are in positions of leadership. Contrary to the common assumption, the situation of women in this respect has not improved over the past few years, and younger women do not appear as leaders in the scientific infrastructures. Of the thirteen women leaders in the Academy only two were about forty years of age, and the rest were older. The requirements for leaders in science are very high: in addition to an adequate formal educational background managerial abilities are also demanded. The advancement opportunities of women are notoriously worse than those of men.

The low rate of women at higher levels of management is an international phenomenon. Although the number of women leaders has increased over the past decades, there is still a marked difference between the respective opportunities of men and women to become a leader. Women are hindered by their family responsibilities, and biases present obstacles to their equal opportunities.

The working day of a leader is defined by the actual work to be done, and it may go well past the normal working hours. In many cases this may cause problems for women. (Koncz, 1987) A number of studies have shown that women – despite their gender-specific traits – are just as able to fill leading positions as men. Unfortunately, the traditional image of a successful leader is based on the personality traits of the male, and this image persists even today. Moreover, men have had ample time over the centuries to adjust to the requirements of leadership while women have only just begun to become familiar with it in the recent decades. Women face a disadvantage because of the persistence of the traditional social division of labour as well as the inadequate support given to the promotion of women's interests. Achievement of the long-term goals of 'woman-policy' may only be possible through creating proper conditions for a general human emancipation, since the emancipation of women is an integral part of the general emancipation.

Conclusions

Although women form an increasingly significant part of the Hungarian scientific community, the majority of women are still facing greater inequalities in their professional roles than their male colleagues. For women this is a recurrent source of frustration, while from the perspective of science policy it leads to an under-utilisation of the research potential. Obstacles to the creativity of women result in a waste of resources available to society. The reasons for the weaknesses in women's scientific creativity are mostly of a socio-economic nature. The preconditions for the improvement of women's position lie in the improvement of their situation on the labour market, at the work place as well as within the family.

References

Adatök a nök hejzeteröl (Data on the position of women). Budapest: Central Office of Statistics, 1982.
Hungarian Academy of Sciences, *Information on its Organisation, Role and Research Network*. Budapest: Institute for Research Organisation, 1988.
KONCZ, K., Nök a munkaeröpiacon (Women on the labour market). KJK, 1987.
SAS, J., Nöies nök és férfias férfiak ('Feminine' women and 'masculine' men). Budapest: Akadémiai Kiadó, 1984.
TAMÁS, P., Kutatónôk a magyar tudományban (Female researchers in the Hungarian science). *Magyar Tudomány*, 2, 1984, pp. 145–55.

11

Stubbornness, Drudgery, Scientific Interests and Profound Commitment[1]

Janni Nielsen and Bente Elkjaer

Introduction

The qualities mentioned in the title are key words given by women in the natural and technological sciences in Denmark when asked why they in particular had succeeded in obtaining a tenured position at a university or other institutions of higher education.[2] They form part of the results of a study on Women in Science and Technology Careers, which was guided by the following question: What obstacles and opportunities did women experience when pursuing a career within natural sciences and technology at universities and other institutions of higher education?

The focus in feminist research on gender and natural sciences and technology is most often on why women are not present in the same number as men. This gender imbalance and the obvious lack of women researchers in natural sciences and technology careers have been addressed by many researchers. The explanatory factors are many, such as the structure and organisation of

---

1. We are greatly indebted to all the secretaries at our institute who helped in numerous ways and to Bo Olsen, a student, who wrote the database programme. The Danish Association for Masters and Ph.D.s secured the anonymity of the study by sending out the questionnaires. Finally associate professor Benny Karpatschof, who is a statistician, commented helpfully on our use of data and tables. Thank you to all.

2. This study does not include women engineers. Included are both lecturers (associate professors) and adjuncts (assistant professors). The latter do not have full security of obtaining a tenured position as their contract of employment only lasts for four years. These categories are referred to here as tenured and semi-tenured positions.

research communities, economic and political factors, gender socialisation and the main responsibility of caring for the family which falls on women, etc. Also the methods and content of science have been critically addressed, and the demarcation of the subjective influence in scientific research has been questioned. Concern has also been expressed with the image – the so-called 'objectivistic illusion' – that science and research suffers from. (Keller, 1985)

However, the interest in the reasons why women are a minority within the natural sciences and technology prevents the search for knowledge about and by the women who are actually there. Thus very little is known about women who actually hold a tenured position in the natural sciences and technology. We lack knowledge of the obstacles – and the opportunities – that women encounter on their way up the ladder of the merit system. This study addresses these aspects. It is an attempt to make the women visible, to let them 'speak' for themselves.

Female Graduates in Tenured Positions

The present study also draws upon an earlier investigation carried out by the Danish Association of Masters and Ph.D.s. (Petersen, 1986) That study showed that to a large extent female graduates live in dual-career families and their partners also have academic careers. Conversely male graduates mainly live with women who have non-academic careers. In general women graduates have fewer children than male graduates, and childlessness is quite common among the younger who are non-tenured or unemployed. The study showed that most of the women holding a full-time tenured position experienced the lack of time as their main conflict. They developed different strategies in order to cope with this lack of time, e.g., they let their children stay longer hours in child care institutions, they sleep less and skip many leisure activities.

The study concludes that it is a handicap to be a woman, in the labour market of academics as well as elsewhere. The life of a female graduate is very much like other women's lives, due to women's position in the labour market and in the family. Equality does not come automatically with a longer education, is the main conclusion.

Design of the Study

The basis for this investigation was a questionnaire sent to women in natural sciences and technology. The criteria for the selection of women in natural sciences and technology careers were the following: (1) employed full-time at universities and institutions of higher education[3] or other governmental research institutions, (2) holding tenured positions (associate professors) or semi-tenured positions (assistant professors) and (3) working in physics, mathematics, statistics, computer science, chemistry, biology, or biochemistry.

The study may be described as retrospective, as the respondents had to answer questions aimed at uncovering their span of life, from graduation to the present. When creating the questionnaire, an epistemological understanding of gender as identity, as symbols and values and as a structural level served as a basis. (Harding, 1986) The question of obstacles and opportunities is approached in two ways, one focusing on the women themselves, and another focusing on the organisation/work conditions. In the questionnaire gender has been used as a descriptive category (marital status, number of children, father's occupation, etc) and as an analytical category (perceptions of organisation, management principles, etc.). We have included both issues that support the gender-stereotype questions and issues that contrast them, as well as factual questions and questions open for interpretation.

The questionnaire was sent to 127 women graduates contacted through the Danish Association of Masters and Ph.D.s, and respondents were anonymous: 66 questionnaires were returned, which gives a return rate of 52 per cent.[4]

Many questions were open-ended, giving the respondent the opportunity to write freely, unrestricted by our expectations. This gave us an abundance of information and generated complexity in the answers. However, the abundance was also a weakness, because in treating the answers to open-ended questions as texts we had to reduce the variety to common themes.

---

3. The job content is defined for all employees as 40 per cent research, 10 per cent administration, and 50 per cent teaching.
4. In order to test representativity we made a trial run and analysed 24 arbitrarily selected questionnaires. The results did not differ significantly from the result of the analysis of all questionnaires received.

The actual treatment was inspired by a phenomenological approach. (Giorgi, 1975) This means reading the text and trying to condense the meaning of the answers through the text itself without using a theoretical framework other than the metatheoretical framework implied in the phenomenological approach.

General Description of the Respondents

The women in the sample are married or cohabiting with university graduates. The spouses also do research in the natural or technical sciences and work at the same type of institutions. One half of the women are between forty and fifty years old. If we include women between forty and fifty-five years of age, the figure is 72 per cent. They had their first child when they were in their late twenties and their second child some years later in their early thirties.

As the entrance exam for their studies, the majority (about 80 per cent) of the women had a General Certificate in mathematics and science. Another 12 per cent either had a linguistic General Certificate supplemented with further entrance examinations, or were granted an exemption from the rule requiring candidates to have passed an entrance exam. A great majority graduated in biology, biochemistry or chemistry. Mathematics and statistics graduates are second, and geography and geology third. Only a small proportion of the respondents studied physics. Almost half of the women have either a Ph.D. and/or a Danish Doctor's degree.[5] The typical career pattern is via scholarship or research fellowship (63 per cent) to a semi-tenured position and then to a tenured position. The majority (90 per cent) are employed as associate professors and only 4 per cent are full professors.

We found it striking that the occupation of about 75 per cent of the respondents' fathers was in technical-administrative areas or in the technological and/or natural sciences. The fathers are far more educated than the mothers. This is not surprising in view of women's general lack of education in the past – and

---

5. The Ph.D. is a research degree obtained at the conclusion of a trainee programme as a researcher. Doctor of Science is a degree obtained on the basis of a comprehensive scientific work by a mature scientist.

even today with respect to higher education. About one-third of the fathers have higher education; for the mothers the number is only nine out of sixty-five, but four out of these nine are in the natural and technical sciences. Surprisingly these four mothers are all married to spouses who hold/held tenured positions in the natural and technical sciences. We also want to draw attention to the fact that the respondents, too, are married to men in the natural and technical sciences. In general the findings may also indicate that one reason for our respondents to choose the technical-natural areas may be the inspiration from their parents, especially the fathers.

Reasons and Motivations for the Choice of Subject

The reason for choosing to study technical or natural sciences was measured by a multiple choice question. The respondents were asked to rank factors of importance for their choice. The choice of study is reportedly mainly influenced by professional interest in their chosen field. Among these, biology and mathematics, but also to a certain degree chemistry are ranked highest. Inspiration from a teacher and a feeling of where one's abilities lie also get high scores, whereas the category of career opportunities and fear of unemployment, as well as inspiration from fathers are ranked low.

The same question, and one dealing with motivation for their choice were asked as open-ended questions. Here professional interests were overwhelmingly emphasised as the reason for choosing the particular field of study. Although considerably less often, a broad interest in nature, humanity and society governed the motivation for the selected study. This is not surprising, considering that a majority of the respondents chose their field of study in the 'softer' end of the 'hard' sciences. The respondents are driven more by interests other than career planning and perhaps even financial interests.

Motivations and Qualifications in Research

In the open answers, the most quoted motivation for research is professional interest. Among the rest of the answers, factors such as saving the earth and nature, interest in people and society are mentioned. These answers may be grouped under the label: care for humanity and society. Wanting to make information technology systems more useful for people in their daily work is another factor that may be included.

Another striking result is that the respondents'motivations for a particular field of research seem to be governed by coincidences that led to possibilities to organise co-operative work with a mentor, colleagues or husband. This may indicate a lack of autonomous personal choice. On the other hand, professional interests may have enhanced the choice.

In an attempt to uncover which qualities respondents find important for a scientist to possess, they were asked to rank three of the most important qualities out of a multiple choice of thirteen. The qualities most often ranked highest are shown in Figure 11.1.[6]

Commitment and curiosity are the most frequently mentioned as the two most important qualities of a scientist. The necessity of a theoretical foundation and ability to organise and structure research work are ranked as the next most important.

Factors Influencing Choice of Research Topic

Women were also asked what they considered to be the most important factors influencing the choice of their research. The five most often ranked factors out of nine possibilities are shown in Figure 11.2.

There is absolutely no doubt about priorities. Personal commitment is most often ranked first, and there is a clear correlation with the women's choice of the qualifications they found most important in a scientist. A theoretical problem, followed closely by a practical problem are next most often ranked first.

Politically decided research priorities, included as one of the nine possible factors influencing research topic selection, are considered of little relevance. This is interesting, in view of the science policy of the last decade, which is marked by increasingly stronger attempts to rule the direction of research in Denmark.

6. The figure shows the seven most often mentioned qualities.

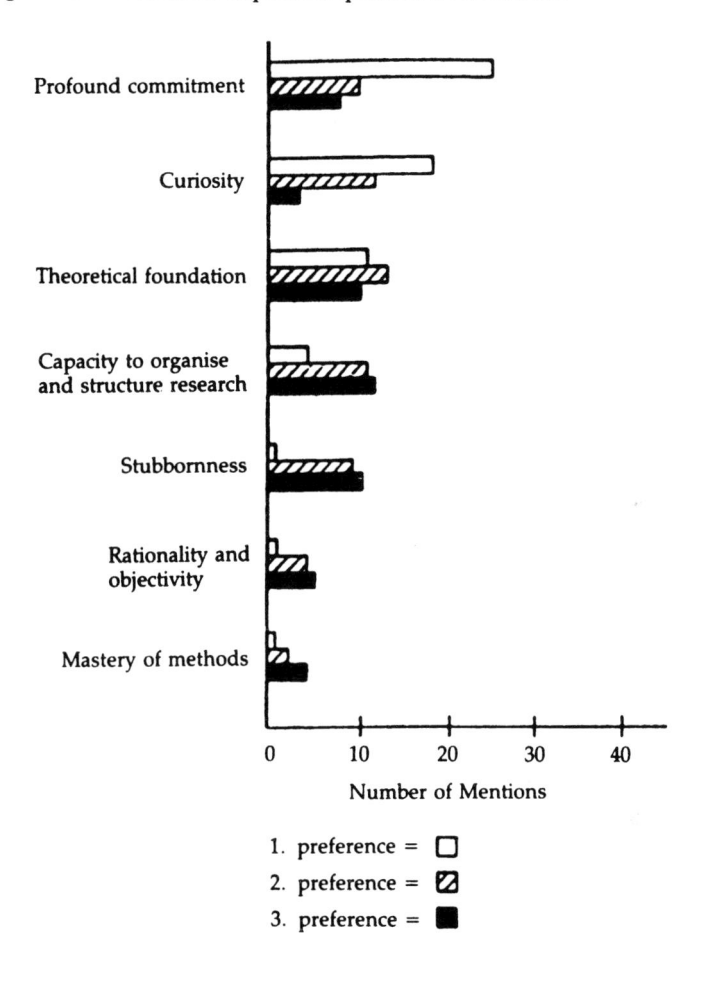

Figure 11.1: The most important qualities in a scientist

[Bar chart showing number of mentions (0-40) across qualities: Profound commitment, Curiosity, Theoretical foundation, Capacity to organise and structure research, Stubbornness, Rationality and objectivity, Mastery of methods]

1. preference = ☐
2. preference = ▨
3. preference = ■

Information was also gathered on the women's research environment. The results show that a considerable part of research is a process where one works alone. However, co-operation with colleagues, both on a national and an international level are also significant characteristics of the environment in which the women work.

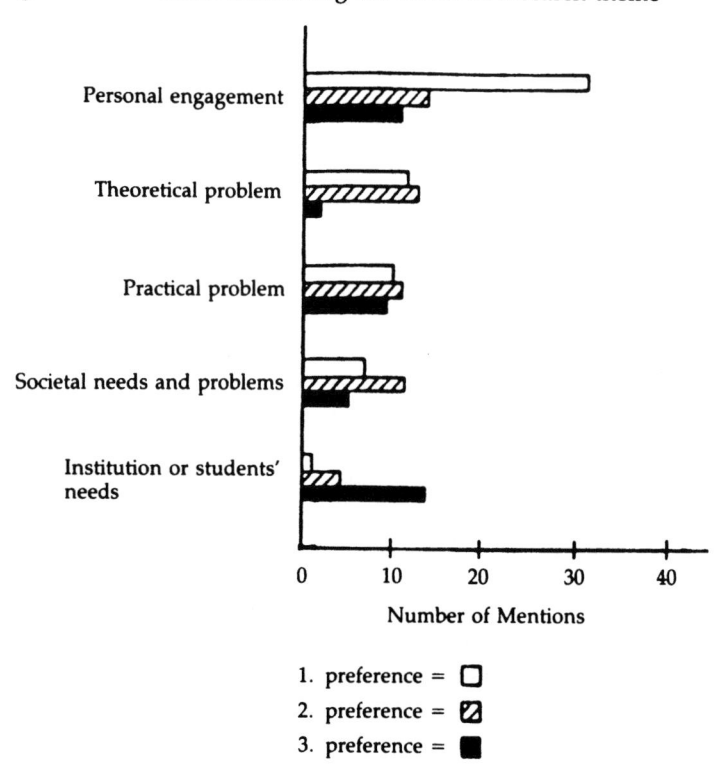

Figure 11.2: Factors influencing the choice of research theme

Personal engagement	
Theoretical problem	
Practical problem	
Societal needs and problems	
Institution or students' needs	

Number of Mentions

1. preference = ☐
2. preference = ▨
3. preference = ■

Reasons for Succeeding within the System

The reasons for succeeding in the system can be divided into three categories. In the first category the focus is on personal qualifications. Such personal properties as stubbornness, followed closely by the capacity for hard work and drudgery are emphasised. But also ambition and professional interests are mentioned as reasons.

The second category includes external factors, or factors outside the respondent's control. Here the most frequently mentioned reasons for having succeeded are good luck or coincidence, and support from family, friends and colleagues. A few women ascribe the reasons for having obtained a tenured position to the possibility of child care, to the postponement of having their first child, or to not having children at all.

Finally in the third category are those who question the whole notion of career and/or ascribe the reasons why they have obtained a tenured position to the historical period when there were few academics and plenty of jobs.

Thus, one may conclude that in order for a woman to succeed in the technical and natural sciences she must be stubborn, work hard and have good luck and support as well as be in the right place at the right time.

Obstacles and Opportunities

The study also aimed at finding out whether women meet any specific obstacles and opportunities in making a career in the technical and natural sciences. It is striking that as many as a fifth of all respondents did not answer to the open question dealing with opportunities. Another fifth answered 'none' or 'none due to gender'. A really striking result is that none of the women ascribe their opportunities to personal qualifications. Instead, the opportunities are ascribed to external or organisational incidents. They regard their specific possibilities as due to the historical period when there were lots of jobs and few people with an academic education. Some regard gender as having been an advantage, and some mention support from husband, friends and colleagues, or absence of a family.

A lot more respondents have answered the open question about obstacles, and a few of them mention 'none' and 'none women specific' obstacles. A third of the women write that the obstacles they have met are due to having children, and a fifth state that they are due to their marriage, the husband's work, or the family. Lack of self-confidence, not being very aggressive, belonging to the school of 'nice girls', showing consideration for other human beings are some of the other answers given. A minority of the answers mention that the women's own perception of their personal qualifications can be a problem – although not the major problem.

Other obstacles mentioned are related to the institution in which the respondents work and to the whole environment of academia. The respondents mention too much teaching and lack of research opportunities, e.g. because they have to stick to normal working hours. They also mention obstacles that are specifically directed towards the fact that they are women – the men on their job do not regard women as one of themselves and moreover, regard women as less clever than themselves. If a woman happens to be just as clever and shows it, she will experience envy from her male colleagues. It is also mentioned that women are expected to participate more actively than men in creating a positive social climate in their work institution.

Attitudes Toward Women in Science

Women scientists' attitudes towards certain allegations about women and science are also of interest. The allegations may be divided into three groups: one concerning qualifications and interests, another concerning perceptions of male scientists, and finally one dealing with the influence of the structure and policies of academic institutions on women's possibilities. The results of the analysis of the respondents' answers are summarised in Figures 11.3, 11.4 and 11.5.

The greatest consensus among respondents prevails on the allegation 'that research demands qualifications and knowledge especially possessed by men': 90 per cent of the women disagree with this. The majority of women (78 per cent) also agree that 'it is easier for a man to make an academic career', and that 'women are more oriented toward the social conditions at their place of work than are their male colleagues' (about two-thirds of the respondents agree). Almost 30 per cent of the respondents neither agree nor disagree with the statement that 'women think of research in a societal/human context', and less than half of them agree with the claim that 'initiatives of equity have made it easier for younger women scientists'.

Discussion

On the basis of the results of this study we may conclude that choices are made at an early age. This is reflected by the fact that the majority of the women's entrance examinations were mathematics and science General Certificates. It is not the eagerness to study physics that characterises most of the respondents, but biology, chemistry and mathematics.

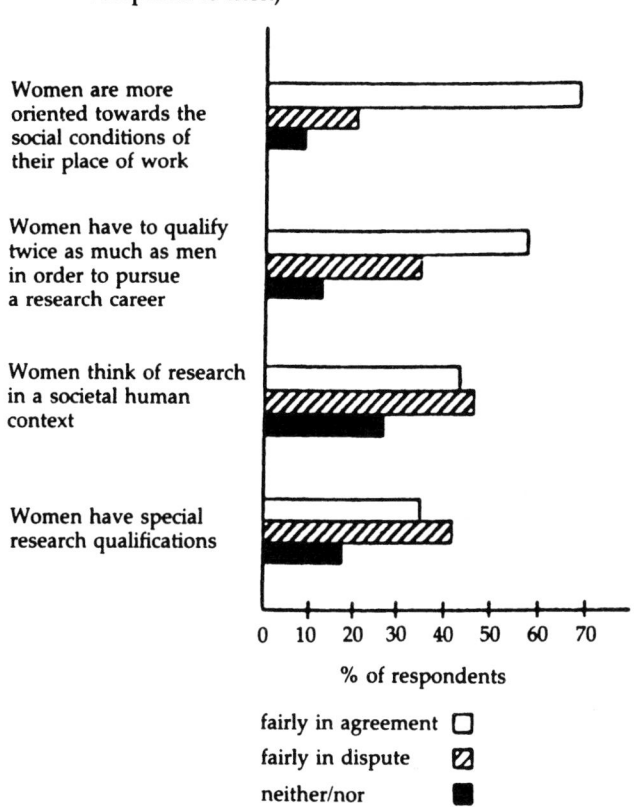

Figure 11.3: Attitudes towards women scientists' qualifications (as compared to men)

- Women are more oriented towards the social conditions of their place of work
- Women have to qualify twice as much as men in order to pursue a research career
- Women think of research in a societal human context
- Women have special research qualifications

% of respondents

fairly in agreement ☐
fairly in dispute ▨
neither/nor ■

The women are highly qualified: almost half of them have a Ph.D. and/or a Danish Doctor's degree. The number of all Ph.D.s in Denmark is not very high. However, it is far more common to have Ph.D.s in the technical and natural sciences than in other fields.[7]

The respondents gave professional interests as the reason for choosing their field of study. However, as many as about 75 per cent of the respondents' fathers' occupation were in fields which to some extent overlapped with those of the respondents. In

---

7. The data are from the Ministry of Education, Economic and Statistics Department.

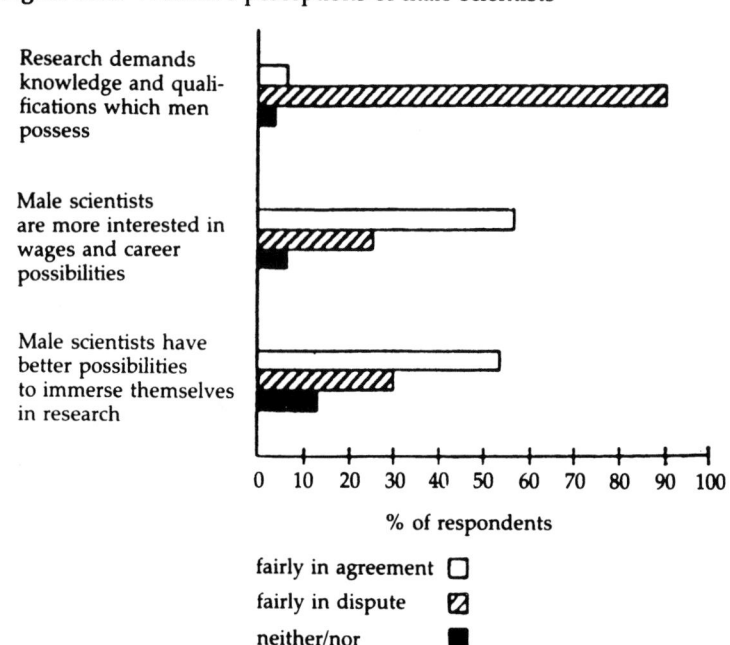

Figure 11.4: Women's perceptions of male scientists

- Research demands knowledge and qualifications which men possess
- Male scientists are more interested in wages and career possibilities
- Male scientists have better possibilities to immerse themselves in research

% of respondents

fairly in agreement ☐
fairly in dispute ▨
neither/nor ■

considering the reasons for 'professional interests' getting such an overall high score we must bear in mind that these women finished their studies a long time ago. Their emphasis on professional interest may be due to the time span and their long personal life-history of dealing with their particular professional interests. Probably in the initial phase also, the fathers' occupation may have had some impact on choosing the natural sciences in high school.

Motivations for doing research showed the same pattern as motivations for studying a particular field. These being – again – professional interests. However, some of the motives given for the choice of research were: care for nature, the environment, human beings and society.

It is interesting that the necessity to think rationally and objectively and of mastering methods is not considered an important qualification in a scientist. This could be interpreted from a feminist perspective to be a reflection of the fact that to be rational, objective and master methods count only within the

Figure 11.5: The influence of the structure and policies of academic institutions

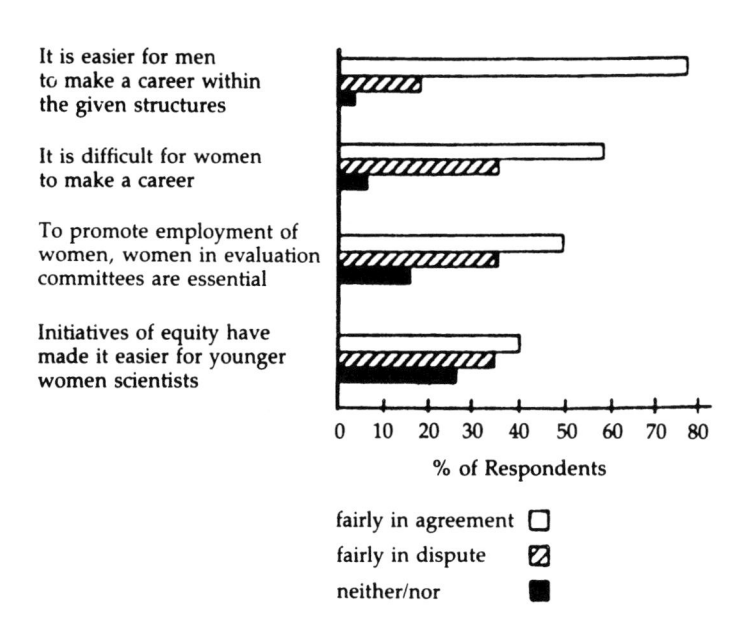

It is easier for men to make a career within the given structures

It is difficult for women to make a career

To promote employment of women, women in evaluation committees are essential

Initiatives of equity have made it easier for younger women scientists

% of Respondents

fairly in agreement ☐
fairly in dispute ▨
neither/nor ■

scientific community ruled by the symbols of masculinity. This interpretation could be supported by the way the women answered: placing value on profound commitment, curiosity and personal interests. However, there is a danger in this argumentation. Firstly, these concepts are disputed and no common agreement exists. Secondly, it is a myth that the scientific process is rational, objective and logical. This myth is – to some extent – fed by the way research results are presented. If research is to be made understandable to others, it has to be communicated. The scientific text must be a fairly straightforward presentation to the reader (it may however be very difficult to read), and as a result the actual scientific process with its interconnected and cross-connected patterns is left out. The scientific process is certainly also very intuitive, and mentally cross-connected. One can suggest that part of this process is reflected in the high ranking of personal commitment and curiosity. This does not mean that the actual research is not governed by the urge to be objective, to create rational arguments and master methods, but research and research processes are also very complicated cognitive and emotional processes that often imply intuition, irrationality, curiosity, etc.

Personal commitment was found to be an important factor influencing the choice of research activities. Moreover, commitment to humanistic and societal needs is, to some degree, inherent in this personal commitment. The low priority assigned to the effect of science policy on research suggests that, above all, personal preference governs the choice and outcome of research. This is not to say that research is governed by merely selfish interests. On the contrary, very few personal commitments are. Scientists are also part of a community, and in most cases they want to choose research topics that currently seem relevant. Besides, they have to apply for funding for their research, equipment, etc. However, one may speculate that official efforts to govern research centrally will have a detrimental effect on the commitment to research, because this demands a very instrumental attitude from the scientist. Research may lose its elements of creativity and become an uncommitted and mechanical process.

Considering obstacles and opportunities, it seems that women do not ascribe their opportunities to being due to their own efforts but to reasons they had no influence over. These reasons are the historical period in which they graduated, good economic background, etc., i.e. external reasons. However, the reasons for obstacles are to some extent internalised. The fact that they also wanted children is seen as the main obstacle. But women also blame themselves for, among other things, lack of self-confidence. This was also found in the study mentioned earlier done of Masters and Ph.D.s. Also in our study women see themselves as being primarily responsible for taking care of the family, and regard this as the most important obstacle to a career.

The family is characterised by emotional relations, and to take care of it demands a certain orientation and qualifications. This orientation and qualifications may be carried over to work. This is why two-thirds of the respondents agree that women are more oriented towards the social conditions at work than men. But they do not think that either women or men possess special qualifications which are essential in research. However, it is clear that women feel that men can better devote themselves to research, because they are not mainly responsible for the family. It is indisputable that the women perceive the family, and especially children as an important obstacle. What one may dispute, however, is why it is so. For the past twenty years of women's studies this factor has been given as an almost undisputable reason for women being the exploited gender, for the existence of a gender division of work, for the lack of public power, etc. Demands for men's participation in taking care of the family, for public child care facilities, for shorter working hours, etc., have been forwarded as solutions for overcoming these obstacles. We do not disagree with these demands, but want to draw attention to the fact that all these demands concern factors outside of women themselves. One could perhaps also benefit from a thorough sociological and psychological study of motherhood, in order to understand the reasons for its importance to women. Motherhood can also be seen as a part of women's bastion of power. We do not want to argue against motherhood as being a work of caring and guided by altruistic motives, just as it may be emotionally deeply rewarding. But, we do want to break the barrier that prevents us from going beyond regarding motherhood and having children as the most important obstacle to a woman's career.

References

GIORGI, A., 'An Application of Phenomenological Method in Psychology', in A Giorgi, et al (eds): *Duquesne Studies in Phenomenological Psychology, II.* Pittsburgh: Duquesne University, 1975.
HARDING, S., *The Science Question in Feminism.* Milton Keynes: Open University Press, 1986.
KELLER, E.F., *Reflections on Gender and Science.* New Haven, Yale University Press, 1985.
PETERSEN, A. STEEN, 'Højtuddannet? – Ja. Ligestilling? – Nej.' (University degree? – Yes. Equity? – No.) Special Issue, *Magisterbladet*, Danish Association of Masters and Ph.D.s, 1986.

12

Is to be an Engineer still a Masculine Career in Spain? Notes on an Ambiguous Change in University Technical Education

Maria Carme Alemany

Introduction

Women increasingly use technology in their daily lives. Domestic appliances, housekeeping gadgets, televisions, video systems and cassette recorders are becoming more frequent in the home and technology is increasingly a part of such activities as driving a car or working with computers. This trend will continue in all technologically developed societies. However, women still remain on the sidelines of these developments. Far from having a creative relation with technology, women are mostly little more than users of certain technological appliances. Daily experience shows that the manufacture, development and application of technology belongs to the 'male world'. There are very few women scientists in the fields most closely related to technology.[1] Women rarely take part in the direct production of goods; only 21.0 per cent of the industrial labour force are women. (Instituto Nacional de Estadística, 1988) By contrast, men start from a dominant position in their relation with technology. It almost goes without saying that technology is perceived as a masculine field.

Nevertheless, most European countries have witnessed an increase in the number of women entering university science

---

1. According to the 1981 Census, in Spain there were only 1,222 women graduates in the field of Engineering and Technology, accounting for 2.3% of all graduates.

Table 12.1: Total number of students enrolled at the Higher Technical School for Engineers in Telecommunication, Polytechnic University of Cataluna

	Men	Women
1981	1587	57
1982	1628	75
1983	1999	87
1984	1956	112
1985	1986	180

and technology programmes since the 1970s. The corresponding rise in women graduates in these fields has tended to reduce the sexual division of professions, although there is still a marked distinction between the humanities, which continue to attract most women, and the masculine domain of scientific and technical studies.

In Spain, the number of women entering higher education has risen since the beginning of the 1970s.[2] This has been accompanied by a relative feminisation of technical studies, with more and more women entering these fields. The number of women enrolled in university technical programmes[3] tripled in the space of ten years, rising from 2,584 in 1975 to 7,044 in 1985, whereas the number of male students has tended to fall since 1982 (see Table 12.2).

What are the causes of this phenomenon? Is technical education no longer a masculine domain? Do careers in technology have the same meaning for both men and women? How is sexist discrimination reflected in these studies? What mechanisms cause this discrimination?

These are the main questions posed in a comparative survey

---

2. The Spanish higher education system comprises three levels: diploma studies (three years), graduate studies ('Licenciatura') (five years), and doctorate (two years). Technical studies also comprise three levels: technical engineering ('Ingeniería Técnica') (three years), advanced engineering ('Ingeniería Superior') (five years plus a further year of project work), and the doctorate (two years). The text refers to graduate studies ('Licenciatura' level) and to advanced engineering ('Ingeniería Superior').

3. In the Spanish system, 'Escuelas Técnicas Superiores' (University Schools of Technology) include studies in Engineering and Technology as well as specialised agricultural science programmes in Agronomy and Forestry.

of students (women and men) enrolled at the 'Escuela Técnica Superior de Ingeniors de Telecomunicaciones' of the Polytechnic University of Cataluna.[4]

Our research combined quantitative analysis of the academic achievements of the students (men and women) over the period 1981–5, with qualitative analysis of students (men and women) in the fourth and fifth years of study. The interviews dealt with their reasons for choosing these studies, the gender relations in the University, and the students' career plans. We sought to locate and analyse those differences between the two groups (men and women) which could account for the difficulty women have in taking up this field of study (see Table 12.1).

The Perverse Effects of Malthusian Policies in University Technical Education

In Spain, as in most countries, university schools of technology practice a Malthusian policy designed to maintain their position and prestige by enabling them to guarantee their graduates high salary levels and a relatively low rate of unemployment. (Consejo de Universidades, 1989) These policies kept increases in the number of students entering these programmes relatively low, even in the midst of a period of general expansion in higher education. In the particular case of the three higher technical schools for engineers in telecommunication in the country, the number of places has barely risen 3 per cent over ten years, despite the current fashionableness of electronics and technologies related with telecommunications (see Tables 12.2 and 12.3).

The fact that there are more candidates than places available in university schools of technology has meant that selection procedures have become increasingly rigorous and rigid, resulting in the exclusion of more and more of those students who have obtained lower grades in the general university entrance exam. This type of selection has probably favoured the admit-

---

4. This research project is part of a comparative study on 'Los comportamientos académicos de los estudiantes en Ingeniería, Física y Filología' (The forms of academic behaviour of the student population in Engineering, Physics and Philology), financed by the Spanish Woman's Institute, The Ministry of Social Affairs.

Table 12.2: Students enrolled at University Schools of Technology in Spain

	1975/6 Total	1975/6 % Women	1985/6 Total	1985/6 % Women
Architecture	13 951	11.8	14 307	22.8
Aeronautical engineering	1 807	1.7	1 727	6.5
Agronomy	2 480	7.2	4 407	22.8
Civil engineering	4 594	1.6	4 517	8.1
Mechanical engineering	16 259	2.2	15 612	7.7
Mining	1 569	2.5	1 596	13.3
Forestry	929	8.6	965	19.2
Naval engineering	1 052	1.8	608	8.2
Electronics and Telecommunications (the three colleges)	5 457	2.4	5 945	8.7
Chemical engineering	567	12.2	403	34.5
Total	48 665	(2 584)	50 087	(7 044)
Total % women		5.3		14.1

Source: Elaborated from University statistics from the Spanish Ministry of Education and Science (1987).

Table 12.3: Total students enrolled in all years of study

	1975/6	1985/6	Growth Rate 1975/85
Non-technical faculties	323 669	550 755	70.2
University Schools of Technology	48 665	50 087	2.9

Source: Elaborated from University statistics from the Spanish Ministry of Education and Science (1987).

tance of women into these programmes, since, although women are a minority in science subjects in secondary schools, their overall grades are higher than those of men. (Centro Nacional de Investigación y Documentación Educativa, 1987) This partly explains why the percentage of men in these programmes has declined since 1982, as other studies have shown. (Marry, n.d.) In this sense, it is possible to speak of the 'perverse' effects of the current model: whatever the original intentions behind the admission policy, the priority given to overall academic criteria has in fact led to a feminisation of the student body enrolled in university technical programmes.

A Career in Technology is not an Asexual Project

In Spain, as in most countries, admission to technical studies at the university level is limited to those students who chose to study sciences in secondary school. Although in principle women can do the same things as men, the different socialisation processes result in there being more boys than girls enrolled in secondary science subjects (52 per cent boys as against 48 per cent girls in the final year of secondary school).

Although the content and orientation of secondary studies are such that students learn about the pure sciences (mathematics, physics, etc.), these studies do not provide any familiarity with technology. Since information on the more technical aspects has thus to be gained outside the education system, it is highly unlikely that both sexes attain the same awareness of technology. As has been pointed out, women are mostly little more than users of technology, and daily experience leads them to associate the actual working of technological machines with the male world. Furthermore it is clear that during childhood the type of games and toys designed especially for boys allow them to become familiar with technical methods, whereas such toys and games are considered unsuitable for girls. Males are thus more likely to develop an aptitude for technical matters.

This hypothesis was fully confirmed by the survey. All the boys interviewed explained that they had played with games like 'Electro-L', 'Microchips-Electronics' and 'electronics kits', or had at least 'pulled apart a lot of radios'. On the other hand, none of the girls said that they had been given toys of this kind, although two did mention that they had played with their brothers' games. Only one woman went counter to this trend: 'I liked electronics a lot when I was a child, and when I was a teenager I was able to practise with my radio-cassette player.'

Boys are introduced to technology within the family environment of their childhood, and this family setting may eventually motivate them to take up a technical career, allowing their parents to experience the corresponding pride of having a 'son in electronic engineering'. Many men explained that their mother or father oriented them towards such careers when they were children.

This combination of different socialising factors makes it clear that men and women approach career decisions from asymmetric perspectives.

It should be pointed out that all but one of the men interviewed said that their decision to take up technical studies at university had been made in the second or third year of secondary school, or even earlier. None of the women had made such an early decision, although they knew that they had a preference for science subjects in their middle years of secondary school. Only a third of the women took their decision during the final year of secondary school, while the rest did not do so until they had finished their secondary education. Most of the women sought additional information before deciding to study electronics, either by approaching a teacher or acquaintance or by attending career guidance talks. In contrast, all the men said that they had been fully aware of the nature of the profession and its technical content, and only two had sought any additional information before deciding on electronics.

Women decided to study electronics at a stage when they only had a vague idea of the subject, without any precise awareness of the actual syllabi or programmes. Men, on the other hand, said that they had found out about the subject through magazines and journals or through contacts with university students in the field.

It should also be noted that the desire to 'become an engineer, fireman or policeman' is part of the men's world. Boys have no trouble identifying with such professions, whereas girls mostly find them contradictory to their own nature, as was indicated when one of the girls admitted that she has 'always liked boyish things'. If the decision to 'become an engineer' is coherent with a male's affirmation of his masculinity and may help him to develop his intellectual creativity, the same decision can still be a 'deviance' for a woman, even in today's world. It is thus clear that the decision to study electronic engineering does not mean the same thing for both sexes.

Many women see the decision to take up technical studies as an 'escape' or a 'break with tradition'. Numerous women indicated that their preference for technical studies was motivated by a desire 'not to teach', since women with degrees in the pure sciences often become teachers in secondary schools.

However, this choice is not always seen as a 'deviance' or 'rupture'. Other women students said that their aim was to 'do something practical' instead of studying the abstract principles and theories of the pure sciences. Considered in isolation, such opinions could be interpreted as a choice for a 'more feminine' alternative and may thus appear to contradict descriptions of the same choice as a 'deviance'. But if one compares the aims of women with those of men, there can be no doubt that the fundamental differences outlined above remain valid.

With respect to men, it is interesting to note some of the reasons behind their inclination for modern technology: 'I've always liked electronics'; 'I wanted to study something up-to date'; 'I didn't want to study anything as specialised as physics or computer science, I wanted something more general'; 'When you study pure science, you have to be a genius before you can do anything interesting; but in electronics there will always be opportunities, even if you're not a great brain'. Mention was also made of better employment prospects and the bright future of technical studies.

On the other hand, only one woman spoke of any passion for electricity and electronics; very few mentioned employment prospects; and those that did added that this aspect had not influenced their decision.

Taken together, the above factors confirm that the women's decision to embark on technical studies is taken under relatively 'opaque' conditions and as a 'transgression of the dominant model'. Even when women do not refer to themselves as deviating from the model, they are still conscious of the ways in which they differ from the men with respect to their initial motivation and study behaviour patterns. This awareness is revealed in statements such as: 'I wanted to do something I liked, but the boys are only interested in getting good jobs'; 'We girls thought about it more carefully before studying here'; 'Even though they don't admit it, the boys are looking for studies with a future'; 'We girls are more motivated'; or 'It's natural for a boy to study electronics, but not for a girl'.

It should be pointed out that none of the men or women interviewed mentioned any family resistance to their choice of studies, although two men said that their parents had been fairly reticent. On the other hand, a third of the women indicated that their choice had surprised their families and that they had been subjected to reminders of what were considered negative aspects for women, including 'the difficulty of the subject', 'living away from home' and 'the lack of women students'. Some of the women who had sought information from their teachers had been advised to take up more feminine studies like the natural sciences, pharmacy or chemistry.

The men indicated that they had always felt supported by their families when they had failed an exam or had to repeat a subject. But this kind of unconditional family support was not confirmed by all the women, some of whom said that their academic problems often gave rise to reminders that 'they should have studied something easier' or to complaints that they are 'making their lives too complicated'. Beyond the family setting, women and men encounter different reactions when they mention their studies in their general social environment. For the men, the reaction is positive, whereas the women encounter reactions of fear or distancing 'because people don't expect you to be studying something like this'.

Overall, these experiences mean that men feel completely at home with technical studies, whereas women take these same studies as a challenge.

Comparisons Between Exam Results

The academic grades of the two groups have been compared on the basis of exam results for each of the subjects taught in the five-year programme over a period of five years (1981–5).[5] The total of 320 examinations with a total of 45,707 marks were tabulated.[6]

The main conclusions to be drawn from the analysis are as follows:

5. This methodology compensates for the small size of the female population in order to allow the analysis of variance to be applied in the comparison of results according to sex.
6. The number of subjects per year is as follows: 1st year: 5 subjects, 2nd year: 6 subjects, 3rd year: 7 subjects, 4th year: 7 subjects, 5th year: 7 subjects.

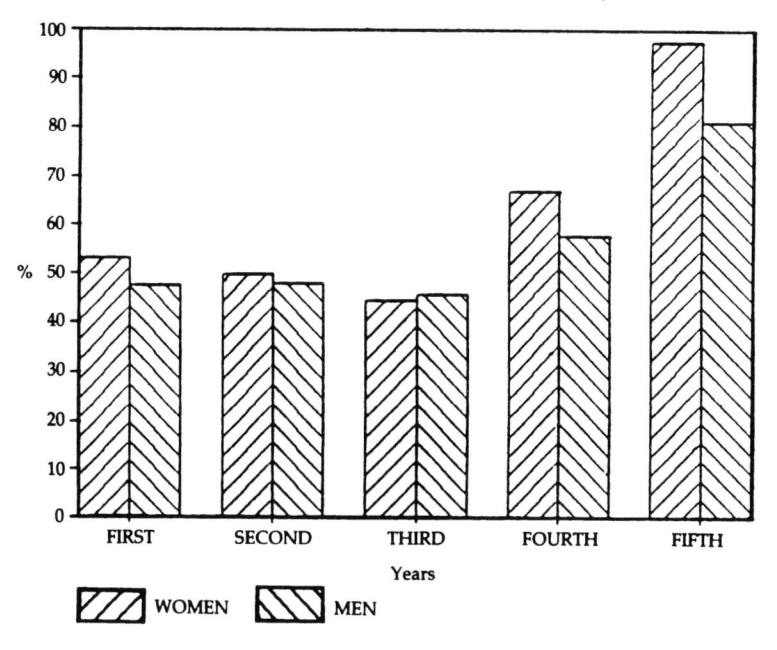

Figure 12.1: Percentage of students having passed each year at the Higher Technical School for Engineers in Telecommunication, Polytechnic University of Cataluna

[Bar chart showing percentages for WOMEN and MEN across FIRST, SECOND, THIRD, FOURTH, FIFTH years]

1. The student attrition rate for both sexes is highest in the first three years and peaks in the third year. This means that the least gifted students drop out and the failure rate is significantly lower in the fourth and fifth years.

2. The women's failure rate[7] is lower in the first, second, fourth and fifth years, and significantly lower in the final two years. This is very probably the result of the less talented women being excluded in the third year (see Figure 12.1).

3. Women have significantly higher grades than the men in the final two years, and when the grades from all five years are analysed as a whole, the women's results are significantly higher.

4. The exam absentee rate is higher for the men in the first,

---

7. The failure rate is taken as the total of fail grades and exam absenteeism over the total of enrolled students.

second, fourth and fifth years. This very probably leads to a higher male drop-out rate, although this latter rate cannot be analysed on the basis of the official exam results.

The lower failure rate for women is probably the result of a group of factors including:

1. The initial admittance criteria favour the entrance of women with better secondary school grades.
2. Discrimination against women taking up university technical studies dissuades many of the less daring women, resulting in a form of self-selection/exclusion (Lavergnas, 1986) by which only those who feel very sure of their own worth decide to take up these studies.
3. The challenge to 'dare' induces forms of survival behaviour manifested in a greater commitment to studies and thus in a higher success rate. From this perspective, good grades act as a way of protecting oneself in a place that is felt to be alien or at least rather 'unfriendly'.

There can be no doubt that success in studies strengthens the position of women in technical studies. But it also increases the selective nature of these studies, leading both to the exclusion of some of the men who would previously have graduated and to the dissuasion of some of the women who are less daring or have less brilliant academic records. The superiority of women also gives rise to direct but veiled competition between the two groups. Women remain unaware of this competitive aspect and are surprised at the negative reactions their success provokes among men: 'I never get angry with the boys just because they might pass when I don't. If anything, I tend to get angry with myself.' In contrast, men take special note of the grades obtained by their women classmates. They justify their particular interest by saying things like: 'Since there are only a few girls, everyone knows their name and you can't help spotting them in the list of exam results.' Men thus recognise and openly admit that the women obtain better grades, adding that 'they're swots', a very common opinion amongst men and a discursive strategy obviously designed to explain away the evidence of women's superiority.

In the same way, fail grades are fully accepted among men, without any negative reflection on the student's intrinsic interest in his future profession. On the other hand, failure among the women easily lends itself to comments about their real interests in choosing the profession, with typical expressions like: 'She's here to catch herself a husband.'

This 'academic harassment' by men leads the women to hide their successes. None of the women interviewed revealed any awareness of their sex achieving better grades. Only after being pressed on this point did they admit that they 'fail less', adding that 'it's something we know but don't talk about'. Their success is then justified by statements like: 'We pass more because we have a better idea of what we want'; 'I wouldn't be studying electronics if I didn't like it a lot'; and 'I knew it would be difficult and I'd have to study hard.'

Discriminatory behaviour thus prevents women from affirming their own group particularity or 'deviance'. Despite the fact that their lower failure rate means that more women are entering technical professions, they tend to take silent refuge behind their academic success. This effectively allows their particularity to be normalised and explained away in terms of the traditionally masculine social and political values typical of technical studies.

This research confirms that, as women, girls in the area of electronics and telecommunications encounter two main problems when beginning their studies: the 'technological illiteracy' resulting from their previous education, and their minority status in a field which is socially considered to be masculine. Both problems present handicaps and mean that girls have to work harder at their studies.

The present research reveals that it is not enough for women to be integrated into this kind of education on a merely formal or statistical basis. Positive measures should also be taken to ensure that girls are on an equal footing with boys.

The research reported on in this paper contributes to a general understanding of the difficulties experienced by women who enter fields which are still socially considered to be 'more suited to men'.

References

Centro Nacional de Investigación y Documentación Educativa (CIDE), *'Mujer y Educación en España'* (Women and education in Spain). Madrid, 1987.

COMINA, C. *'Athéne, Rôle et importance des jeunes filles dans les Ecoles d'ingénieurs en Europe* (On the role and importance of women in engineering universities in Europe). Lausanne: SEFI, 1983.

Consejo de Universidades – Secretaría General, 'El stock de titulados universitarios y su relación con el mercado de trabajo 1976–86' (The present supply of higher education graduates in relation to the needs of economic and social development in Spain). Madrid, 1989.

INE (Instituto Nacional de Estadística): *Encuesta de Población Activa. Tercer trimestre 1988.* (Survey on the active population. Third quarter 1988).

LAVERGNAS, I., 'Le corps étranger ou la place des femmes dans l'institution scientifique' (The foreign body or the place of women in research institutions). Montreal, 1986. (Unpublished Ph.D. thesis of Sociology).

MARRY, C., 'La Féminisation de la catégorie d'ingénieurs en France: importance, limites et causes d'un phénomène récent' (The feminisation of the category 'engineers' in France: the importance, limits and causes of a recent phenomenon) Paris: CNRS, n.d. (in print).

Ministerio de Educación y Ciencia. Estadísticas Universitarias (University Statistics). Madrid, 1985 and 1987.

13

Recommendations

Dorothea Gaudart

The United Nations system has finally become aware of the need to promote women's participation as actors in all aspects of sciences and technology including in policy-making and decision-making posts. The international meeting of experts (Dubrovnik, December 1984) convened by UNESCO's Sector for Social and Human Sciences recommended that obstacles encountered by women in careers of natural, physical and social sciences and opportunities available to them be examined in countries of different regions. This recommendation served as an impetus for the studies undertaken by social scientists from the European region and presented in this publication. Research along similar lines may be useful in other countries and other regions and can help provide some guidelines to the scientific community and policy-makers in moving towards gender equity in the sciences.

In 1985, 157 states committed themselves to taking concrete measures to eliminate all forms of sex-based discrimination by the year 2000 by adopting the Nairobi Forward-looking Strategies for the Advancement of Women.[1] These strategies emphasise that 'the full and effective participation of women in the decision-making and implementation process related to science and technology, including planning and setting priorities for research and development, and in the choice, acquisition, adaptation, innovation and application of science and technology for

---

1. The Nairobi Forward-looking Strategies (hereafter FLS) were adopted by United Nations General Assembly Resolution A/40/108. FLS in: *Report of the World Conference to review and appraise the achievements of the United Nations Decade for Women: Equality, Development and Peace,* Nairobi, Kenya, 15–26 July 1985, A/Conf.116/28/Rev.1, N.Y., United Nations: 1986. Paragraphs 1–372, pp. 2–89.

development should be enhanced (paragraph 200)'.[2] Subsequently, the United Nations (1990–95) system-wide medium-term plan for women and development requests 'to collect data on the stock of women scientists and technologists and those active in their fields by age groups and by discipline'. (UN, 1987)

UNESCO for its part has continued to give some attention to the situation and needs of women scientists and engineers, notably within its *Third Medium Term Plan, 1990 to 1995*. This time Unesco's Sector for Natural Sciences is requested to promote some 'activities to overcome the obstacles to women's access to scientific and technological training and careers and increase women's involvement in science and technology in co-operation with national authorities and professional organizations'.[3]

Participation of women in R & D provides an input for the implementation of these legitimated objectives. Such input is particularly important for improving and raising the standards of the scientific and technological culture, and it will add another dimension to it. In this context, certain generalisations and recommendations for policies can be deduced from the findings of the national studies on women in positions of responsibility in R & D presented in this volume.

Some policy recommendations relate to the educational system, many more to the unsolved reconciliation of professional and family obligations including the related imbalanced time budgets, and the majority to the necessary structural changes in

---

2. This paragraph 200 of FLS orginates from a resolution on 'Women, science and technology' adopted by the United Nations Conference on Science and Technology for Development (Vienna, August 1979). It continues with the invitation of Member States to facilitate: 'The equal access for women and men to scientific and technological training and to the respective professional careers'. A/CONF.81/L.4/Rev.1.

3. There seems to be remarkable progress in UNESCO's programming. The international scientific community is especially addressed in the Third Medium Term Plan, within Major Programme Area II 'Science for Progress and the Environment'. The last paragraph in Resolution 25 C/4/102 related to this Major Programme Area II invites the Director-General to pay special attention, during the biennial programming for 1990–95, '3.(f) to ways and means of strengthening participation by women in all the activities of Major Programme Area II and particularly the training of women specialists'. Another provision in this resolution, as in that on the transverse theme – *Women*, concerns the 'broadening of the co-operation with competent institutions of the United Nations system and other intergovernmental organizations and with the international scientific community through the competent international non-governmental organizations, particularly the International Council of Scientific Unions'.

the scientific community itself at the national, regional and international levels.

Changes in the educational systems at all levels are progressing rather slowly, although the core of the educational problems of girls was recognised nearly forty years ago (Unesco, 1956, 1968, 1980, 1985; Unesco/CEPES, 1988; Gaudart 1975; Sutherland, 1985; Whyte, 1984). Before the turn of the millenium, efforts should be strengthened. For example, curricula in technical and vocational education should be renewed to emphasise the design and use of new information and communication technologies. In textbook development the biographies and professional achievements of women scientists and inventors should be included.[4] Within the framework of the 1990 International Literacy Year and its follow-up, a campaign should be launched to promote the 'technical literacy of girls and women' by the year 2000.

Careful research is needed into the images of engineering and technology projected in the public eye, on the one hand, and social sciences and humanities, on the other, taking into account both sexes. Today, engineering and technology seem not only to be viewed as being a contradiction to humanities, but also to be rejected as a field of study on account of social and personal concerns about environmental problems. Scientific interests of women are interrelated with a profound commitment to humanity and society. All envisaged programmes on scientific progress and the environment are equally important for women and men. Therefore, they should fully integrate women as contributors to, and beneficiaries of, scientific and technical co-operation and, consequently, in educational programmes concerning environmental issues.

Continuing higher education and post-graduate training are particularly important for women who frequently have to interrupt their education or professional careers for childbearing and childrearing. However, in view of the social changes in the

4. Examples of the increasing literature, not only by women: Vare, E.A., Ptacek, G., *Patente Frauen, Große Erfinderinnen*, (Mothers of invention). Wien/Darmstadt: Zsolnay, 1989; Moussa, F., *Les femmes inventeurs existent, Je les ai rencontrées* (Women inventors exist, I met them). Genève, 1986; Feyl, R., *Der lautlose Aufbruch, Frauen in der Wissenschaft* (The silent start, Women in science). Berlin: Neues Leben, 1981: *Frauenstudium und akademische Frauenarbeit in Österreich* (Women's higher education and academic work in Austria) 1968–1987. Vienna: Austrian Federation of University Women, 1987.

status of women, on the one hand, and changes related to the accelerating scientific progress, on the other, particularly in the field of biotechnology, it seems essential to include university teachers in projects of continuing education for the twenty-first century.

The traditional 'division of labour both at home and at the work place' persists in the science sector. The fact that scientific maturity coincides with marriage and motherhood has not yet been challenged even by the highly qualified scientists of the East and West. Obviously the understanding that 'the role of both parents in the family and in the upbringing of children. . . . requires a sharing of responsibility between men and women and society as a whole' has not yet been put into practice.[5] This lack of consciousness of both women and men in academic circles towards gender equity in science might be interpreted as one of the major obstacles towards scientific careers of European women. The lack of such consciousness is reinforced by the strong feelings of responsibility women harbour for home-life and child care. On the other hand, women scientists who have children consider the childraising period to be the least productive period of their careers.

Time management evokes a number of questions. Time for research work could vary but domestic and parental work remains constant, it was stated. However, both women and men scientists need time for the acquisition of socially important information about opportunities for specialisation abroad, outside the intrascientific institution, grants, vacancies, fundraising, etc. There is room for further research on time budgets

5. *The Convention on the Elimination of All Forms of Discrimination against Women*, adopted by the United Nations General Assembly came into effect in September 1981. By 1 December 1989, 99 States had become parties to the Convention, having either ratified or acceded to it (in the European Region 29 State Parties; not yet signed: Albania, Malta; signed but not yet ratified: Netherlands, Switzerland, United States of America). In the European Region too, the participating States of the Conference on Security and Co-operation in Europe 'confirm their determination to ensure equal rights of men and women. Accordingly, they will take all measures necessary, including legislative measures, to promote equally effective participation of men and women in political, economic, social and cultural life. They will consider the possibility of acceding to the Convention on the Elimination of All Forms of Discrimination against Women, if they have not yet done so.' Principle 15 of the *Concluding Document of the Vienna Meeting 1986 of Representatives of the Participating States of the Conference on Security and Co-operation in Europe, held on the basis of the Provisions of the Final Act Relating to the Follow-up to the Conference.*

Recommendations

and changing gender relations aiming at sharing domestic, parental and family responsibilities and reconciling them with professional obligations and scientific interests. In view of the endogamy in the professions in general, and therefore also in the science sector, more research is needed on two-career couples in science.

Generally speaking, women experience greater difficulty than men in finding suitable employment, and it is also more difficult for them to reach positions of responsibility. Conversely, the status accorded to women scientists has an important influence on their choice of fields of science and technology. It is a vicious circle. Women are, more often than not, obliged to accept jobs for which they are overqualified. This should create concern about wasting human resources within the scientific community of the European region.[6]

The situation may well change in the future with the increase in the number of women attending and graduating from universities, and their growing inclination towards scientific and technical fields, which in the public opinion of the old continent of Europe is still held to be the preserve of men.

As far as career prospects are concerned, studies confirm that the personnel policy in the public sector leaves less room for discrimination than in the private sector. (ILO, 1977) In many instances, therefore, recommendations and strategies for changing this situation are first and foremost the responsibility of the public authorities. (Unesco, 1985; Sutherland, 1985; Unesco/CEPES, 1988)[7]

---

6. The evolution of the careers of women upon completion of higher education, graduate unemployment and the difficulties encountered by women in promoting their careers and in attaining leading positions in the professions they exercise were subjects of discussion at the European Symposium on *The Role of Women in Higher Education, in Research, and in the Planning and Administration of Education* (Bucharest, 17–19 October 1988), organised jointly by the Division of Equality of Educational Opportunity and Special Programmes of UNESCO (ED/SPO) and the European Centre for Higher Education (CEPES). It was recommended to urge UNESCO, CEPES and the Office of Statistics to collect existing information on the situation of women higher education graduates and to undertake new steps for providing new data concerning the transition period between completion of university studies and integration into the labour market. Furthermore UNESCO should co-operate closely with the ILO as well with the various international governmental and non-governmental organisations which carry on activities related to the role of women in higher education in particular and in society in general.

7. In a wider perspective it might be recalled that the 1974 UNESCO Rec-

Some countries and institutions have moved forward with equity strategies or systematic programmes, such as positive action programmes or special temporary measures, in order to accelerate de facto equality between women and men in scientific activities. For example, where both a woman and a man applicant present themselves, men should not be chosen automatically, but women should receive – as a temporary special measure – preferential treatment.[8] Similar kinds of programmes exist in the context of the European Community as 'positive action'[9] or in North America as 'affirmative action'. (Vogel-Polsky 1985, 1989) Accordingly, such personnel policy options at the level of the research institutions or companies can improve the position of women scientists, as indicated in some of the case studies.

Furthermore, women with the appropriate qualifications should be given opportunities in sectors in which they have not usually been employed. They should be encouraged to hold positions of responsibility while those with higher scientific attainments should be provided with the possibility of pursuing more challenging research tasks. Women scientists should be explicitly included in the exchange of scientists, in the development of a scientific and technological culture as well as in the popularisation of science and technology.

Women themselves have a fundamental role to play in bringing about changes to improve their situation. It is only by showing themselves prepared to occupy positions of responsibility and fit to do so that their employers or institutions will

---

ommendation on the Status of Scientific Researchers contains non-discrimination clauses, *inter alia*, on sex. However, this recommendation might need some up-dating in view of equal opportunities or the above-mentioned Convention on the Elimination of all Forms of Discrimination against Women, particularly its Article 4.

8. Article 4 of the UN Convention on the Elimination of All Forms of Discrimination against Women states: '1. Adoption by States Parties of temporary special measures aimed at accelerating de facto equality between men and women shall not be considered discrimination as defined in the present Convention, but shall in no way entail as a consequence the maintenance of unequal or separate standards; these measures shall be discontinued when the objectives of equality of opportunity and treatment have been achieved. 2. Adoption by States Parties of special measures, including those measures contained in the present Convention, aimed at protecting maternity shall not be considered discriminatory.'

9. Council Recommendation on the promotion of positive action for women, 84/635/EEC of 13 December 1984.

Recommendations

adopt a personnel policy of the type described above. It cannot be emphasised too strongly that women already holding positions of authority (few though as they still are) can have a decisive influence in this respect. Only by the increasing involvement of women in the public affairs of the scientific community can the effectiveness of the full utilisation of the R & D personnel be improved.

Summing up the national studies, there has been an undeniable increase in the recruitment of women scientists in recent years. However, there is need for research on terms and conditions of employment of women scientists, on the one hand, and on requirements and qualifications needed to occupy positions of responsibilities in R & D, on the other. More research is needed, in particular, on the practices of the scientific community itself. For example, how do women fare as applicants, rank and file scientists, project leaders, expert referees (peer review-system) or members of research policy bodies? Furthermore, age limits for applications or grants should be reviewed as to whether they are designed to fit scientists with uninterrupted professional careers and thus implicitly exclude women scientists. Nominations, designations, appointments and election to executive boards of scientific bodies as well as the underlying regulations or legislation should be included in these endeavours in order to consider adding supplementary provisions or criteria to distribute scientific responsibilities more equitably between men and women.[10]

To this end, UNESCO should be invited to investigate whether the various international scientific associations and learned societies, scientific councils, networks etc. in their statutes and budget programmes include any provision on equal opportunity or any kind of temporary special measures to accelerate de facto gender equity. UNESCO's support of these competent international non-governmental organisations, such as the International Council of Scientific Unions (ICSU) and the International Social Science Council (ISSC), should be contingent on the existence of such measures.

---

10. See also recommendations of UNESCO International meeting of experts on factors influencing women's access to decision-making roles in political, economic and scientific life and on measures that may be taken to increase their responsibilities (Dubrovnik, 10–14 December 1984), paras 183, 184, 186.

Positive or affirmative action policy or temporary special measures to increase the number of women in positions of responsibility are also needed in UNESCO's Secretariat itself to ensure that women scientists play an equitable role in the execution of their programmes.

UNESCO should be further invited to request its Member States to support, develop, monitor and periodically evaluate science and technology policy in favour of gender equity in R & D.[11] The impact of gender equity policy on women in R & D should be monitored and evaluated periodically, at regional and international levels. UNESCO might envisage a European or world conference on these topics in order to exchange experiences gained. The comparative findings should be widely disseminated so as to encourage more women to pursue scientific careers and more men to open the doors for women scientists' access to the competent national and international scientific communities.

References

Austrian Federation of University Women, *Frauenstudium und akademische Frauenarbeit in Österreich* (Women's higher education and academic work in Austria) 1968–1987. Vienna, 1987.
FEYL, R., *Der lautlose Aufbruch, Frauen in der Wissenschaft* (The silent start, women in science). Berlin: Neues Leben, 1987.
GAUDART, D., 'Zugang von Mädchen und Frauen zu technischen Berufen' (Access of girls and women to technical careers), in A. Niegl, (ed.) *Series on the Education of Girls and Women*, vol. 3. Vienna: Österreichischer Bundesverlag, 1975.
ILO, *Conditions of work and employment of professional workers*, Tripartite Meeting on Conditions of Work and Employment of Professional Workers. Geneva 1977.
MOUSSA, F., *Les femmes inventeurs existent, je les ai rencontrées* (Women inventors exist, I met them). Geneva, 1986.
Unesco, *Access of Girls and Women to Technical and Vocational Education,*

---

11. *National science and technology policies in Europe and North America, Present Situation and Future Prospects*. Paris: Unesco, 1978, includes a section on 'Access of women to careers and policy-making' in the National Summaries which deal with the human resources engaged in science and technology.

Recommendations

Preliminary Study jointly prepared by Unesco and ILO for the 10th session of the United Nations Commission on the Status of Women, 1956.

——, *Comparative Study on Access of Girls and Women to Technical and Vocational Education* (ED/MD/3). Paris, 1968.

——, *International congress on the situation of women in technical and vocational education* (ED-80/CONF.401/5). Bonn, 1980.

——, *International seminar on opening up to women of vocational training and jobs traditionally occupied by men* (ED/80/CONF.708/4). Frankfurt, 1980.

——, *International meeting of experts on factors influencing women's access to decision-making roles in political, economic, and scientific life and on measures that may be taken to increase their responsibilities* (Dubrovnik, 10–14 December 1984).

——, *International meeting of experts on 'Reflection on women's problems in research and higher education'* (Lisbon, 17–20 September 1985). Final Report, 1985, (SHS/CONF/85/612/16).

——, *National Science and Technology Policies in Europe and North America, Present Situation and Future Prospects*. Paris, 1978.

——, Working Document: *Cross-cultural research project on women's participation in positions of responsibility in careers of sciences and technology*, by C. Marias, Division of Human Rights and Peace, prepared for the Unesco/Vienna Centre preparatory meeting (Zagreb, Yugoslavia, 21–23 July 1988) for the project on this theme.

——/CEPES, European Symposium on 'The role of women in higher education, in research and in the planning and administration of education', (Bucharest, 17–19 October 1988). Final Report, 1988.

United Nations, *System-wide Medium Term Plan for Women and Development for the Period 1990–1995*, Economic and Social Council Document E/1987/52.

SUTHERLAND, M.B., 'The Role of Women in Higher Education', *Higher Education in Europe*, 10:3, 1985, pp. 47–56.

VARE, E.A. and G. PTACEK, *Patente Frauen, Große Erfinderinnen* (Mothers of invention). Vienna/Darmstadt: Zsolnay, 1989.

VOGEL-POLSKY, E., 'Positive action programmes for women', *International Labour Review*, vol. 124, no 3, 1985, pp. 253–65; vol. 124, no 4, 1985, pp. 385–99, Geneva, 1985.

——, *Les Actions positives et les contraintes constitutionnelles et législatives qui pèsent sur leur mise en œuvre dans les Etats membres du Conseil de l'Europe*. Strasbourg: Conseil de l'Europe, EG (89) 1, 1989.

WHYTE, J., *Encouraging girls into science and technology: some European initiatives*, Unesco (Science and technology education document series, Number 7, ED-84/WS/64 Rev.). Paris, 1984.

14

Select Bibliography: Women in Scientific and Technical Careers*

Ruža Fürst-Dilić

ABIR-AM, P.G., AND D. OUTRAM, eds, *Uneasy Careers and Intimate Lives – Women in Science, 1789–1979*. New Brunswick, N.J.: Rutger University Press, 1989.

ACAR, F., 'Turkish Women in Academia: Roles and Careers.' *METU Studies in Development* (Ankara), vol. 10, no. 4, 1983, pp. 409–46.

ACAR, F., 'Role Priorities and Career Patterns of Women in Academe: A Cross-Cultural Study of Turkish and Jordanian University Teachers', in S. Lie and V. O'Leary (eds), *Storming the Tower: Women in the Academic World*. New York: Kogan Page (forthcoming).

ACFAS, 'Etre femme de science'. *Actes du Colloque*, University of Quebec I, May 1985.

ACKERMANN, W., 'Cultural Values and Social Choice of Technology', *International Social Science Journal*. Paris, Unesco, vol. 33, no. 3, 1981, pp. 447–65.

AISENBERG, N. AND M., HARRINGTON, *Women of Academe: Outsiders in the Sacred Grove*. Amherst: University of Massachusetts Press, 1988.

AMRAM, F., 'A Woman's Work Includes Invention', *The Woman Engineer*, vol. 1, no. 4, 1981, pp. 29–35.

ANDERSON, M.R., 'Women in Science and Engineering: A Case of Awareness and Encouragement', *First International GASAT Conference*, Eindhoven, 1981, pp. 1–18.

ARNOT, M., *Race and Gender: Equal Opportunities Policies in Education*. London: Pergamon and Open University, 1985.

AVERY, D., 'Home Versus Job: A Global Perspective on Women in Science'. *Unesco Features*, no. 796, 1984, pp. 12–17.

BARAD, J., 'A Process for Career Decision-making', *The Woman Engineer*, vol. 1, no. 4, 1981, pp. 35–7.

BERANEK, W. JR. AND G. RANIS, (eds), *Science, Technology and Economic Development: A Historical and Comparative Study*. New York: Praeger, 1978.

BERGHAHN, S. et al. (eds.), *Wider die Natur? Frauen in Naturwissenschaft*

und Technik. Berlin: Elefanten Press Verlag, 1984.
Bibliographic Guide to Studies on the Status of Women. Development & Population Trends. Paris: Unesco, 1983.
BILLING, Y.D. AND A. BRUVIK-HANSEN, 'Women's Attitude to Technological Studies (Especially Engineering)'. *Second International GASAT Conference*. Oslo, 1983, pp. 175–8.
BIP, *Ingénieur, un métier féminin? Représentation comparée des filles et des garçons en première année d'étude supérieure quant à leur avenir social et professionnel*. Report for Unesco. 1985, 2 vols.
BLEIER, R., *Science and Gender*. New York: Pergamon Press, 1984.
BRADFORD, J., 'Women Scientists in New Zealand. Why so Few?' *Impact of Science on Society*, vol. 30, no. 1, January-March 1980, pp. 37–42.
BRUER, J.T., 'Women in Science: Toward Equitable Participation', *Science, Technology and Human Values*, vol. 9, no. 3, pp. 3–7.
BYRME, E., *Women and Engineering: A Comparative Overview*. Queensland: University of Brisbane, 1984.
CACHELOU, J., 'De Marie Curie aux ingénieurs de l'an 2000. Quatre générations de femmes-ingénieurs', *Culture Technique*, no. 12, March 1984, pp. 265–71.
CHABAUD-RYCHTER, D., 'Technologie et division sexuelle du travail, rapport de synthèse', *Cahiers de l'APRE No. 7*. Paris: C.N.R.S., vols. 1 and 3, April-May 1988.
CHAKRAVARTHY, R., A. CHAWLA, AND G. MEHTA, 'Women Scientists at Work – An international study of six countries', *Scientometrics*, vol. 14, nos. 1–2, 1988, pp. 43–74.
CHATON, J.H., *Etude sur l'accès, en France, des femmes à l'enseignement et à la formation scientifiques et aux carrières correspondantes*. Paris and Geneva: Unesco & ILO, 1981.
CHETVERIKOV, V.N., 'Women's Participation in Higher Education in the USSR', *Higher Education in Europe*, vol. 6, no. 3, 1981, pp. 35–9.
———, 'La participation des femmes dans l'enseignement supérieur en URSS', *Enseignement supérieur en Europe* (Unesco/CEPES), July-September 1981, pp. 39–43.
CHIVERS, G.E., 'Loughborough University Women in Engineering Project'. *First International GASAT Conference*, Eindhoven, 1981, pp. 55–70.
———, 'Women Engineers: A Career Opinion Survey', *Electronics and Power*, May 1977, pp. 370–73.
———, 'Developing a Programme to Assist Girls and Women to Study and Pursue Careers in Science, Technology and Engineering', *Report to the European Institute of Education*, 1982/3, 1983.
———, 'A Comparative International Study of Intervention Strategies to Reduce Girls' Disadvantages in Science and Technology Education and Vocational Training', *International Symposium on Interests in Sci-*

ence and Technological Education, 1984.

———, 'Report on Visits to Sweden in Connection with Comparative Research on Initiatives in European Countries to Encourage More Girls and Women to Study and Take up Careers in Science and Technology', *ESRC Research, Visits Scheme for Social Scientists, 1983–4*, 1984.

———, VAN MENTS, M., *Women in Technology – Proceedings of a Conference held at Loughborough University, January 1984*. Loughborough, 1984.

CLEGG, A. AND W. DUNCAN, 'Girls and Science in Botswana', *Third International GASAT Conference*, London, 1985, pp. 151–9.

CNPF, 'Madame l'Ingénieur', *Revue des entreprises*, no. 436, June 1982. pp. 58–60.

COCKBURN, C., *Machinery of Dominance: Women, men and technical knowhow*. London: Pluto Press, 1986.

COLE, J.R. AND COLE, S., *Social Stratification in Science*. Chicago: University of Chicago Press, 1981.

COLE, J.R., 'Women in science', in D.N. Jacobson and Z.P. Rushton, (eds), *Scientific Excellence, Origins and Assessment*. London: Sage, 1987, pp. 359–75.

DEN BANDT, M.L., 'Gelijke kansen voor akademisch gevormde vrouwen? Vergeet het maar' (Female Opportunities for Women in Academia? Forget it!). *Wetenshap en samenleving*, no. 9, 1975.

D'ONOFRIO-FLORES, P. AND S.M. PFAFFLIN, (eds), *Scientific-Technological Change and the Role of Women in Development*. Boulder, Colo. Westview Press, 1982.

DUBOIS, E.C. et al, *Feminist Scholarship: Kindling in the Groves of Academe*. Urbana and Chicago: University of Illinois Press, 1985.

DURCHOLZ, P., 'The Hidden Career: How Students Choose Engineering', *Engineering Education*, April 1979, pp. 718–22.

———, 'Women in a Man's World: The Female Engineers', *Engineering Education*, January 1977, pp. 292–9.

DURIO, H.F. AND C.A. KILDOW, 'The Non-retention of Capable Women Engineering Students', *Research in Higher Education*, vol. 13, no. 1, 1981.

EASLEA, B., *Science and Sexual Oppression: Patriarchy's Confrontation with Woman and Nature*. London: Weidenfeld & Nicolson, 1980.

EITB; SMITH, J.C., *Women in Engineering – Some International Comparisons*. March 1984.

EKEHAMMAR, B., 'Women and Research: Perceptions, Attitudes and Choice in Sweden', *Higher Education*, vol. 14, 1985, pp. 693–721.

ELIAS, N., 'Scientific Establishments'. In N. Elias, H. Martins and R. Whitley (eds), *Scientific Establishments and Hierarchies*. Dordrecht: D. Reidel, 1982, pp. 3–70.

ELGQVIST-SALTZMAN, I., 'Educational Reforms – Women's Life Patterns:

a Swedish Case Study', *Higher Education*, vol. 17, 1988, pp. 491–504.
EPSTEIN, C. AND R.L. COSER, *Access to Power: Cross-National Studies of Women and Elites*. London: Allen and Unwin, 1981.
Equal Opportunities Commission, *Report on Women into Science and Engineering*. UN, 1984.
ERICKSON, L.R., 'Women in Engineering: Attitudes, Motivations and Experiences', *Engineering Education*, November 1981, pp. 180–2.
European Coordination Centre for Research and Documentation in Social Sciences. *The Changing Role of Women in Society. A Documentation of Current Research*. Research Projects in Progress 1981–3. W. Richter, H. Hogeweg-de Haart and L. Kiuzadjan (eds), with an introduction by Herta Kuhrig. Berlin: Akademie-Verlag, 1985.
——, *The Changing Role of Women in Society. A Documentation of Current Research*. Research Projects in Progress 1984–7. W. Richter, L. Husu and A. Marks (eds). Berlin: Akademie-Verlag, 1989.
FARNHAM, C. (ed.), *The Impact of Feminist Research in the Academe*. Bloomington: Indiana University Press, 1988.
FARLEY, J. (ed.), *Sex Discrimination in Higher Education: Strategies for Equality*. Ithaca: New York State School of Industrial and Labor Relations, 1981.
——, *Academic Women and Employment Discrimination: A Critical Annotated Bibliography*. Ithaca: New York State School of Industrial and Labor Relations, 1982.
GARKE, E., 'Women and Higher Education in Switzerland: Access to Universities, Study Preferences and Job Opportunities', *Higher Education in Europe*, vol. 6, no. 3, 1981, pp. 26–32.
GASTON, J., (ed.), *The Sociology of Science*. San Francisco: Jossey-Bass, 1978.
GAUDART, D., 'Die Frau als Chef: Vorurteile werden revidiert' (Woman as Boss: Prejudices Revised), in *Frauen in der Industrie*, ÖIAG Journal, no. 1, 1983.
GREENGLASS, E.R., 'The Psychology of Women and Scientific Research', in D.N. Jacobson and J.P. Rushton (eds), *Scientific Excellence, Origins and Assessment*. London: Sage, 1987, pp. 341–57.
GRIFFITHS, D., 'The Exclusion of Women from Technology', in W. Faulkner and E. Arnold (eds), *Smothered by Invention*. London: Pluto Press, 1985.
——, AND E. SARAGA, 'Would more Women Change Science?' *Physics Bulletin*, vol. 31, 1980, pp. 166–7.
GORDON, L., Y. GRUZDEVA AND E. KLOPOV, *Women in the World of Science and Technology, Expansion of the Participation of Women in Scientific-Technological and Industrial Occupations: Experience in the USSR and the European Socialist Countries*. 1985.
GORNICK, V., *Women in Science: Portraits from a World in Transition*. New

Select Bibliography

York: Simon and Schuster, 1983.
HAAS, V.B. AND PERRUCCI, C.C. (eds), *Women in the Scientific and Engineering Professions*. Ann Arbor: University of Michigan Press, 1984.
HAAVIO-MANNILA, E. AND I. ESKOLA, 'The Careers of Professional Women and Men in Finland', *Acta Sociologica*, vol. 18, nos. 2–3, 1975.
HAAVIO-MANNILA, E. 'The Position of Women', in Allardt, E. *et al.*, (eds) *Nordic Democracy*. Copenhagen: Det Danske Selskab, 1981.
HARDING, J. 'How the World Attracts Girls to Science', *New Scientist*, 15 September 1983, pp. 754–5.
——, *Science and Technology – A Future for Women?* Paris: Unesco, 1985.
HARDING, S., *The Science Question in Feminism*. Milton Keynes: Open University Press, 1986.
—— AND J.F. O'BARR, (eds), *Sex and Scientific Inquiry*. Chicago, University of Chicago Press, 1987.
HÄYRYNEN, L., 'Naiset akateemisilla urilla' (Women's Academic Careers), in L. Husu, M.-L. Honkasalo (eds), 'Työ nainen ja tutkimus' (Work, Women and Research). Helsinki, Valtioneuvoston kanslian monisteita, 1984.
HEETER, M.C., *Access of Women and Girls to Science Education and to Associated Technical Careers: A Bibliography*. Paris: Unesco, 1980.
HENNING, M. AND A. JARDIN, *The Managerial Woman*. London: Pan, 1979.
HINTON, K. (ed.), *Women in Science*. London: Siscon/Butterworth, 1977.
HORN, H., 'Women in the Israeli Universities', *The Role of Women in Higher Education* (Unesco/CEPES), vol. 1, September 1985.
JENSEN, K., 'Women's Work and Academic Culture: Adaptations and Confrontation', *Higher Education*, vol. 11, 1982, pp. 67–83.
KELLER, E.F., *Reflections on Gender and Science*. New Haven: Yale University Press, 1985.
KELLY, A. (ed.), *The Missing Half: Girls and Science Education*. Manchester: Manchester University Press, 1981.
KOHLSTEDT, G.S., 'In from the Periphery: American Women in Science, 1830–1880', *Signs*, vol. 4, no. 1, 1978, pp. 81–96.
KONCZ, K., 'The Historical Process of Feminization in Intellectual Professions in Hungary from 1980 to 1980', in E. Vamos (ed.), *Women in Sciences: Options and Access*. Budapest: The National Museum of Science and Technology, 1987, pp. 136–59.
KVANDE, E., 'Deviants or Conformists? On Female Engineering Students and Work-related Values and Attitudes', *Second International GASAT Conference*, Oslo, 1983, pp. 189–98.
LANTZ, A., 'Women Engineers: Critical Mass, Social Rapport and Satisfaction', *Engineering Education*, April 1982, pp. 731–7.
LEMOINE, W., 'Women's Place in Science in Venezuela (1875–1958)', *Unesco Community Service Review* (UCS), nos. 85–6, pp. 55–9.

LIE, S., 'Girls and Science and Technology', *Second International GASAT Conference*, Oslo, 1983.
LIHOCKY, J., *Study on Access of Women to Specialized and Scientific Education and Training and to Corresponding Careers in the Czechoslovak Socialist Republic*. Paris and Geneva: Unesco & ILO, 1980. p. 82.
LONSDALE, K., 'Women in Science: Reminiscences and Reflections', *Impact of Science on Society*, vol. 25, 1975, pp. 147-52.
LUUKKONEN-GRONOW, T. AND V. STOLTE-HEISKANEN, 'Myths and Realities of Role Incompatibility of Women Scientists', *Acta Sociologica*, vol. 26, nos. 3-4, 1983, pp. 267-80.
MARENTIC-POZARNIK, B., 'The Neglected Half of Humanity', in *The New Millenium: Women Facing Change. Scientific and Technological Aspects. Final Report*. Ljubljana, 25-9 August 1986. Ljubljana, Unesco International Centre for Chemical Studies of the Edvard Kardelj University, 1986, pp. 164-8.
MARRY, C., 'Femmes ingénieurs: une (ir)résistible ascension?', *Information sur les Sciences Sociales*, vol. 28, no. 2, 1989, pp. 291-344.
MARTIN, B.R. AND J. FROINE, 'Women in Science – The Astronomical Brain Drain', *Women's Studies International Forum*, vol. 5, 1982, pp. 41-68.
MARTIN, P.Y., D. HARRISON, and D. DINITTO, 'Advancement for Women in Hierarchical Organisations: A Multilevel Analysis of Problems and Prospects', *Journal of Applied Behavioural Science*, vol. 19, no. 1, 1983, pp. 19-33.
DE MEURON-LANDALT, M., 'How a Woman Scientist Deals Professionally with Men?', *Impact of Science on Society*, vol. 25, 1975, pp. 147-52.
MICHEL, J., 'Women in Engineering Education', *Studies in Engineering Education 12*. Paris: Unesco, 1988.
MCNUTT, A., 'ET's Message to Women: Break Out of the Mold', *Engineering Education*, May 1983, pp. 805-7.
(The) New Millenium: Women Facing Change, Scientific and Technological Aspects, Final Report. Ljubljana, Unesco International Centre for Chemical Studies, Edvard Kardelj University, 1986.
NEWTON, P., 'Deciding on Engineering: Implications of a Nontraditional Career Choice', *EOR Bulletin*, no. 7, 1983, pp. 74-83.
OTT, M.D., 'Retention of Men and Women Engineering Students', *Research in Higher Education*, vol. 9, 1978, pp. 137-50.
Onderzoek naar Wetenschap, Technologie en Samenleving (Research in Science, Technology and Society), W. v. Rossum, E.K. Hicks and J. v. Eijndhoven (eds). Amsterdam: SISWO publication no. 326, 1987.
PINCH, T. and W. BIJKER, 'The Social Construction of Facts and Artefacts: Or How the Sociology of Science and the Sociology of Technology Might Benefit Each Other', *Social Studies of Science*, vol. 14, 1984, pp. 399-41.

Select Bibliography

POLLARD, L.A., *Women on College and University Faculties: A Historical Survey and a Study of Their Present Academic Status*. New York: Amo Press, 1977.

PRPIC, K., *Marginalne grupe u znanosti* (Marginal Groups in Science). Zagreb: Institut za drustvena istrazivanja Sveucilista u Zagrebu, 1987.

——, 'Zena u znanosti' (Women in Science), *Zena* (The Woman), vol. 40, no. 4, 1982, pp. 53–67.

——, Socijalni profil istrazivaca (A Social Profile of Researchers), *Nase teme* (Our Themes), vol. 29, nos. 1–3, 1985, pp. 141–61.

——, N. HARITASH AND R. CHAKRAVARTHY, *Bibliography on Women in Science and Technology from the Asian and Pacific Region*. New Delhi: National Institute of Science, Technology and Development Studies, 1985.

——, N. HARITASH AND R. CHAKRAVARTHY, (eds), *Qualitative and Quantitative Data on Women in Science and Technology in Asia and the Pacific Region*. New Delhi: National Institute of Science, Technology and Development Studies, 1985.

RÁTY, T., L. SZANTO AND T. SZANTO, 'The Training, Qualification and Mobility of Scientific Manpower', in K.O. and L. Pal, (eds), 'Science and Technology Policies in Finland and Hungary'. Budapest: Akademiai Kiado, 1985, pp. 141–164.

REMY, D. *La problématique de la participation des chercheurs féminins aux activités de recherche*. Louvain: Université Catholique de Louvain, Institut des Sciences Politique et Sociales, 1977.

Research and Social Goals (theme), *Impact of Science on Society*, vol. 29, no. 3, 1979.

RESKIN, B.F., 'Sex and Status Attainment in Science', *American Sociological Review*, vol. 41, no. 4, 1976, pp. 597–612.

——, 'Sex Differentiation and the Social Organization of Sciences', in Gaston, F. (ed.), *The Sociology of Science*. San Francisco: Jossey-Bass, 1978.

Responsabilité des femmes dans la conduite de leur carrière et enseignement supérieur. Rapport préparatoire (Laufer, J. Le cas des femmes d'entreprise; Delavault, H. Les cadres de l'Enseignement et de la recherche scientifique). Table Ronde Unesco-FIFDv Document de Travail, Unesco, March 1988, pp. 23–36.

ROSE, H., 'Beyond Masculinist Realities: a Feminist Epistemology for the Sciences', in Bleier, R. (ed.), *Feminist Approaches to Science*. New York: Pergamon Press, 1986, pp. 67–76.

ROSSER, S.V. (ed.), *Feminism within the Science & Health Care Professions: Overcoming Resistance*. New York: Pergamon Press, 1988.

ROSSITER, M.W., 'Sexual Segregation in the Sciences. Some Data and a Model', *Signs*, vol. 4, 1978, pp. 146–51.

243

RUDNICK, D.T. AND S.E.D. KIRKPATRICK, 'Male and Female E.T. Students: A Comparison', *Engineering Education*, May 1981, pp. 765-70.
RUDNICK, D.T. AND WALLACH, E.J., 'Encouraging Women into Technical Careers: One Successful Approach', *Engineering Education*, May 1979, pp. 802-4.
RUIVO, B., 'A mulher e o poder professional. A mulher em actividades de investigacao cientifica em Portugal' (Women and Professional Power: Women in Scientific Research in Portugal), *Análise Social*, vol. 22, nos. 92-3, 1986, pp. 669-80.
——, 'The Intellectual Labour Market in Developed and Developing Countries: Women's Representation in Scientific Research', *International Journal of Science Education*, vol. 9, no. 3, Summer 1987, pp. 385-91.
RUNDNAGEL, R., 'Frauen in Naturwissenschaft und Technik', *Das Argument*, no. 155, 1986, pp. 74-85.
SANDI, A.-M., 'Science, Development and Women's Emancipation', *Impact of Science on Society*, vol. 30, no. 1, 1980, pp. 53-60.
SANSANWAL, D.N., 'Sex Differences in Attitude to Interest in and Achievements in Science. A Review of Indian Researches', *Second International GASAT Conference*, Oslo, 1983, pp. 151-62.
SAYERS, J., 'Science, Sexual Differences, and Feminism', in B. Hess and M. Ferree (eds), *Analysing Gender. A Handbook of Social Science Research*. London: Sage, 1987, pp. 68-91.
SCHIEBINGER, L., 'The History and Philosophy of Women in Science: A Review Essay', in S. Harding and J.F. O'Barr (eds), *Sex and Scientific Inquiry*. Chicago: The University of Chicago Press, 1987, pp. 7-34.
SCHMIDT, S., *Ingenieurwissenschaften und Naturwissenschaften – Arbeitsmarkt und Nachwuchs in der Bundesrepublik Deutschland und in Bayern*. December 1981.
Science in Policy and Policy for Science (theme). *International Social Science Journal*, vol. 28, no. 1, 1976
Science, Technology and Women: A World Perspective. Washington D.C.: American Association of the Advancement of Science, 1985.
SEFI, 'Women in Engineering Education' *European Journal of Engineering Education*, vol. 11, no. 3, 1986, pp. 231-350.
——, 'Women Challenge Technology', *European Conference on Women, Natural Sciences and Technology*, Elsinore, November 1986.
——, Comina, C., *Athène: rôle et importance des jeunes filles dans les ecoles d'ingénieurs en Europe*. Lausanne, 1983.
Seminar on Women and Technological Innovation; Report. Paris, Unesco and Women's Studies Unit, University of Pertanian, Malaysia, June 1987.
SESSAR-KARPP, E., 'The Situation of Women in Technical Professions in the Federal Republic of Germany', *European Journal of Engineering Education*, vol. 11, no. 3, 1986, pp. 315-20.

Select Bibliography

SHAPLEY, D., 'Obstacles to Women in Science', *Impact of Science on Society*, vol. 25, no. 2, 1975, pp. 115–24.

SIMEONE, A., *Academic Women: Working Towards Equality*. South Hadley, Mass.: Bergin and Garvey, 1987.

SKOG, B., 'Curriculum Options: A Barrier Against Women's Participation in Scientific and Technological Work?', *Second International GASAT Conference*, Oslo, 1983.

DE SOLE, G. AND L. HOFFMANN, (eds), *Rocking the Boat: Academic Women and Academic Processes*. New York, Modern Language Association, 1981.

Some Ideas from Women Technicians in Small Countries, *Impact of Science on Society*, vol. 30, no. 1, 1980.

SPENDER, D. (ed.), *Men's Studies Modified: The Impact of Feminism on the Academic Disciplines*. Oxford: Pergamon, 1981.

STOLTE-HEISKANEN, V., *Women's Participation in Positions of Responsibility in Careers of Science and Technology: Obstacles and Opportunities*. University of Tampere, Department of Sociology and Social Psychology, Working Papers no. 26, 1988.

——, 'The Role and Status of Women Scientific Research Workers in Research Groups', *Research in the Interweave of Social Roles: Jobs and Families*, JAI Press, vol. 3, 1983, pp. 59–87.

——, 'Kobiety i nauka: sprawa plici w tworjeniu wiedzy', *Zagadnienia naukoznawstwa*, no. 1, 1985, pp. 65–77.

Study on Access of Women to Science Education and Training and Associated Careers in Malaysia. Paris and Geneva: Unesco & ILO, 1983.

SULLEROT, E. (ed.), *Le fait féminin*. Paris: Fayard, 1978.

——, 'The Promotion of Women in the World of Work in Europe', *The Courier* (Unesco), November 1978, pp. 18–22.

Technology and Cultural Values (theme), *International Social Science Journal*, vol. 33, no. 3, 1981, pp. 431–521.

TEKELI, S., A. GUNALP, and A. PAYASLIOGLU, Status of Women in Higher Education in Turkey', *The Role of Women in Higher Education* (Unesco/CEPES), vol. 1, September 1985.

TERBORG, J.R. and D.R. ILGEN, 'A Theoretical Approach to Sex Discrimination in Traditionally Masculine Occupations', *Organisational Behaviour and Human Performance*, vol. 13, no. 3, 1975, pp. 352–76.

TUANA, N. (ed.), 'Feminism and Science, I & II', *Hypatia*, vol. 2, no. 3, 1987; vol. 3, no. 1, 1988.

TURI Z.F., 'Women Technical Graduates in Hungary', *Impact of Science on Society*, vol. 30, no. 1, 1980, pp. 23–32.

TUSI, L., 'Women's Scientific Creativity', *Impact of Science on Society*, vol. 25, no. 2, 1975, pp. 105–14.

Unesco, *National Science and Technology Policies in Europe and North America 1978. Present Situation and Future Prospects*, Science Policy

Studies and Documents no. 43. Paris: Unesco, 1979.

——, *Participation of Women in R & D – A Statistical Study*. Paris: Unesco, Division of Statistics on Science and Technology, Office of Statistics, September 1980. (CSR-S-9).

——, *Female Participation in Higher Education. Enrolment Trends, 1975–1982*. Paris: Unesco, Division of Statistics on Education, Office of Statistics, February 1985, (CSR-E-50).

——, *International Meeting of Experts on Factors Influencing Women's Access to Decision-Making Roles in Political, Economic and Scientific Life and on Measures that May Be Taken to Increase their Participation* (Dubrovnik, Yugoslavia, 10–14 December 1984). *Final Report*. Paris: Unesco, 1985.

——, *The Status of Women. Annotated Bibliography for the Period 1965–1985*. Paris: Unesco, 1986. (BEP/87/WS/3).

——, CEPES. *Higher Education and Economic Development in Europe, 1975–1980*. 2 vols.. Paris: Unesco/CEPES, 1983.

US *Woman-Engineer* (Society of Women Engineers), October, vol. 30, no. 2, 1983.

VETTER, B.M., 'Women Scientists and Engineers; Trends in Participation', *Science*, no. 214, 1981, pp. 1313–21.

——, E.L. BABCO, AND J.E. MCINTIRE, (eds), *Professional Women and Minorities – A Manpower Data Resource Service*. Washington: Scientific Manpower Commission, 1978.

VIGLIETTA, M.L., 'How do Women Succeed in Studying Science and Technology? The Situation in an Italian University', *Second International GASAT Conference*, Oslo, 1983, pp. 1–6.

WHYTE, J., 'Encouraging Girls into Science and Technology: Some European Initiatives', in *Science and Technology Education Documentation Series 7*. Paris: Unesco, 1984.

WISTRAND, B., *Swedish Women on the Move*. Stockholm, 1981.

Women Engineer at Work, *Industrial Engineer*, vol. 8, no. 4, 1976, pp. 24–5.

Women in Academe, *Women's Studies International Forum* 6, no. 2, 1983 (Special Issue).

Women in Higher Education (theme), *Higher Education in Europe*, vol. 6, no. 3, 1981, pp. 5–43.

Women in Science: A Man's World (theme), *Impact of Science on Society*, vol. 25, no. 2, 1975.

Women into Science and Engineering, *New Civil Engineer*, 12 January 1984, pp. 13–16

Women, Science and Society, *Signs*, vol. 4, no. 1, Autumn 1978 (Special issue).

Women in Higher Education (theme), vol. 6, no. 3, 1981, pp. 5–43.

Women and Minorities in Science and Engineering. Washington D.C.: National Science Foundation, 1984.

ZIMMERMANN, J. (ed.), *The Technological Woman*. New York: Praeger, 1983.

ZUCKERMAN, H., 'Persistence and Change in the Careers of Men and Women Scientists and Engineers: A Review of Current Research', in L.S. Dix (ed.), *Women: their Underrepresentation and Career Differentials in Science and Engineering. Proceedings of a Workshop*. Washington, D.C.: National Academy Press, 1987, pp. 123–56.

* This select bibliography on women in careers in science comprises primarily books, chapters in the books and articles in periodicals published during the last decade or so in European countries and/or in relation to Europe. However, some representative examples from other continents are taken into account as well.

Notes on Contributors

Feride Acar has a Ph.D. in Sociology. She is currently Associate Professor of Political Sociology. Her areas of academic interest are social and political change, social movements and women's studies. Her ongoing research includes studies of academic women and women in Islamic fundamentalist movements.
Address: Middle East Technical University, Department of Public Administration, Inönu Bulvari, Ankara 06531, Turkey.

Maria Carme Alemany is a specialist in the Sociology of Work, and the Sociology of Education, Human Resources and Education Policy and Reform. She did her main research work in these fields focusing upon the situation of women. She is currently Director of the Centre for Studies on Women and Society (CEDIS).
Address: Centre for Studies on Women and Society, Muntaner, 178, 5°, 1a, 08036 Barcelona, Spain.

Nora Ananieva holds a Ph.D. in Philosophy and is Professor of Political Sciences. Her main works deal with comparative constitutional law and the comparative analysis of political systems, problems of representative government and the role of women. She served as Deputy Director of the Institute of Studies for Contemporary Social Sciences of the Bulgarian Academy of Sciences, and is now Chairperson of the Department of International Relations at the Economic University of Sofia. She is a member of Parliament and its Deputy Prime Minister.
Address: Complex Mladost', Bloc 221, BC 1199, Sofia, Bulgaria.

Marina Blagojević holds a M.S. in Sociology and is currently assistant at the University of Belgrade. Her research work covers social demography, sociology of the family, women's studies, migration and marginal groups.
Address: University of Beograd, Faculty of Philosophy, Department of Sociology, Cika Ljubina 18–20, 11000 Beograd, SFR Yugoslavia.

Ann R. Cacoullos holds a Ph.D. in Philosophy. She is currently Associate Professor of William Paterson College at the State University of New Jersey and Associate Professor at the University of Athens. She is the author of a book on human rights, and of articles on ethics and social and political philosophy. She recently published a monograph on Greek women in technology.
Address: The William Paterson College of New Jersey, Department of Philosophy, Wayne, N.J. 07470, USA; and/or: University of Athens, Department of Mathematics, Panepistemiopolis Conponia, Athens, Greece.

Notes on Contributors

Bente Elkjaer holds a Ph.D. in Educational Studies. Her main research interest is in the field of information technology with an educational and organisational approach where gender is an important perspective. She is currently Associate Professor.
Address: Copenhagen Business School, Informatics and Management Accounting, Howitzvej 60, 2000 Frederiksberg, Denmark.

Ruža Fürst-Dilić holds a M.S. in Sociology and Social Anthropology. Her main research interest is in the fields of women's studies, family sociology, youth sociology, rural sociology, and social demography. Currently she is Scientific Secretary seconded at the Vienna Centre from Yugoslavia.
Address: European Coordination Centre for Research and Documentation in Social Sciences, Grünangergasse 2, 1010 Vienna, Austria.

Dorothea Gaudart holds a Ph.D. in Psychology and is Professor of Empirical Social Sciences and Occupational Sociology at the Institute of Sociology of the University of Vienna. Her main teaching and research is on labour relations, with an emphasis on social and occupational positions of women. She is a member of the Executive Board of the Austrian National Commission for Unesco, where she is chairing its special committee on the status of women. Currently she holds the post of Director of the Division on Labour Relations and Working Women's Affairs at the Austrian Federal Ministry of Labour and Social Affairs.
Address: Federal Ministry of Labour and Social Affairs, Stubenring 1, 1010 Vienna, Austria.

Agnes Haraszthy holds a Ph.D. and currently is a Senior Research Associate at the Institute for Research Organisation of the Hungarian Academy of Sciences. Her main scientific interest is in the fields of sociology of science, evaluation of scientific activity, research organisation and science policy.
Address: Institute for Research Organisation of the Hungarian Academy of Sciences, Münnich Ferenc utca 18, 1051 Budapest, Hungary.

Esther K. Hicks holds a (European) Doctorate in Interdisciplinary Social Sciences and a Dr. S. in Theoretical Anthropology. She has done archaeological and anthropological field work, has led field study courses and has lectured at the departments of Anthropology (medical, cultural and physical anthropology), History and Sociology. Currently she is Head of the Science Research Department at the Netherlands Universities' Social Research Centre.
Address: The Netherlands Universities' Social Research Centre (SISWO), P.O. Box 19079, 1000 GB, Amsterdam, the Netherlands.

Vitalina Koval holds a Ph.D. in Economics. Currently she is a Senior Researcher at the Institute of World Economy and International Relations of the USSR Academy of Sciences. Her main interest is in labour movement studies, introduction of new technology and working women.
Address: Institute of International Labour Movement of the USSR Academy of Sciences, Kolpachnyi perelok 9a, 101831 Moscow, USSR.

Janni Nielsen holds a Ph.D. in Educational Psychology and Informatics. Her main research is in human-computer interaction, cognitive qualifications and learning processes in relation to the development, implementation and evaluation of technological information systems. She is now holding the post of lecturer.

Address: Copenhagen Business School, Informatics and Management Accounting, Howitzvej 60, 2000 Frederiksberg, Denmark.

Heidrun Radtke holds a Ph.D. in Philosophy. She specialises in the fields of personality theory, sociology of work and research on women, and has published work, especially in the field of personality development under conditions of scientific-technological progress. She is currently Head of the Research Group 'Woman' at the Institute of Sociology and Social Policy of the GDR Academy of Sciences.

Address: Institute of Sociology and Social Policy, Otto-Nuschke-Str. 10/11, 1086 Berlin, Germany.

Veronica Stolte-Heiskanen holds a Dr. SSc. and is Professor of Sociology. Her main research interest is in sociology of science. Over the past twenty years she published extensively in the field of science, and science and technology policy studies. Her research and publications also include studies on the problems of women in science. Currently she is Vice Rector of the University of Tampere and is a member of the Finnish National Commission for Unesco.

Address: University of Tampere, Department of Sociology and Social Psychology, P.O. Box 607, 33101 Tampere, Finland.

Annexe A: Other Specialists Who Participated in Various Stages of the Project

Eva Bartova	Institute of Philosophy, Czechoslovak Academy of Sciences, Prague, Czechoslovakia
Jennifer Cassingena	National Council of Science and Technology, Valetta, Malta
Anna Maria Ginerva Conti Odorisio	Inter-University Centre for Women's Studies, Rome, Italy
Inga Eigquist-Saltzmann	Department of Education, University of Umeå, Umeå, Sweden
Maria Lado	Research Institute of Labour, Budapest, Hungary
Gunilla Leander	Department of Education University of Umeå, Umeå Sweden
Beatriz Ruivo	National Board for Scientific and Technological Research (JNICT), Lisbon, Portugal
Maryanne Spiteri	Senglea, Malta
Barbara Tryfan	Rural Sociology Centre, Polish Academy of Sciences, Institute of Rural and Agricultural Development, Warsaw, Poland
Stephen Mills	Assistant Secretary-General, International Social Science Council, Paris, France
Carrie Marias	Programme Specialist, Division of Human Rights and Peace, UNESCO, Paris, France
Willem Stamatiou	Publications and Information, European Coordination Centre for Research and Documentation in Social Sciences, Vienna, Austria

Index

ability (prejudices concerning), 69
academic
   institutions (structure), 208, 211
   positions, 82–5, 102–4, 155–8
   status, 152–5
   utilisation of women, 69–70
Academic Preparatory Education, 176
Academic of Sciences, 4
   Austria, 14
   Bulgaria, 98–9, 102, 107, 108
   Finland, 39, 41, 49, 57–8
   Hungary, 195–8
   Soviet, 119–20, 122–3, 125–9, 131–2
Acar, E., 150–1, 154–5, 160, 165, 167–8
achievement, expectation and, 89–90
administrative posts, 160, 162–7
advancement (factor analysis), 108–16
affirmative action, 232, 234
age factor, 195, 196, 208
agriculture (Bulgaria), 97, 102
Ananieva, N., 95
ASOEE programme, 138
Atatürk, Mustafa Kemal, 150
attitudes towards women, 1, 5, 208–9
Austria (research and development), 9, 31
   dual career couples, 28–30
   education system, 10–11
   employment trends, 11–20
   research activities related to women, 23–6
   research personnel policy, 26–7
   research policy, 21–3

Beekes, A., 173, 182
Bergman, S., 44, 46, 56
Blagoev, D., 95
Blagojević, M., 80

Bourdieu, P., 46, 47, 51
Bruun, K., 47
Bulgaria, 95, 115–16
   historical background, 96–9
   scientific hierarchy, 100–8
   women's advancement, 108–14

careers, 144–5, 231
   choice (educational factors), 176–8
   science and technology, 199–213
   see also role conflict
Cataluna Polytechnic University, 216–17, 223
Cattell, 28
CEPES, 229, 231
children
   childcare, 65, 90–1, 112, 132, 168, 229–30
   childbearing, 7, 45, 77
   protection of (USSR), 120, 121–2
civil service (Austria), 26–7
Cole, J., 42, 44, 46
commitment, 212
committee membership, 50, 54–5
costs, rewards and, 90–93

de Hemptinne, Y., 28
De Jong, 180, 181
decision-making, 28, 107–8, 227
   Finland, 47–55
degree students, 102, 104–6, 127–8, 153
Dekkers, H., 176
Delft Technical University, 179–80, 181
democratisation, 119–20, 128–9, 132
Denmark, 199
   attitudes towards women, 208
   discussion, 208–13

Index

female graduates, 200
 motivations and qualifications, 203–4
 obstacles and opportunities, 207–8
 research topic (choice), 204–6
 study design/respondents, 201–3
 subject choice, 203
 success within system, 206–7
Dickinson, J.P., 28
Diem-Wille, G., 29
Dimitrov, D., 95
Dimitrov, K.R., 98
Dinkova, M., 97
Directorate Council (Bulgaria), 107
Disco, C., 36
discrimination, 154–5, 164–5, 224–5, 231
division of labour, 5, 45, 197, 230
doctoral students, 102, 105–6
Dohnal Johanna, 26
Doorne-Huiskes, J. van, 182
double-burden scenario, 135, 145
dual career couples, 28–30, 200, 231
DVSV, 181

education, 2–3, 228–9
 Austria, 10–11
 Bulgaria, 96, 97–8
 Finland, 36–8, 55, 56
 Germany, 65–6
 Greece, 136–42
 Netherlands, 176–9
 Soviet, 120–1
 Spain, 215–25
 Yugoslavia, 80–2
Eiduson, 28
election of women, 107–8, 128–30
Eliou, M., 139
elitism, 76, 79, 142, 150–1, 155–6
emancipation programmes, 183–4, 186
employment, 183
 responsible posts, 66–70
 structure (USSR), 125–8
 trends (Austria), 11–20
 see also careers; labour market
Engels, Friedrich, 95
engineering, 65–6, 215–25
equal opportunities, 114–15, 123–5, 193–8
Erkut, S., 150, 169
examination results, 222–5
expectation, achievement and, 89–90

failure rate (exams), 223, 224, 225
family, 70, 89, 96
 Bulgarian case study, 110, 111, 112
 division of labour, 5, 45, 197, 230
 dual-career, 28–30, 200, 231
 socialisation, 93, 151, 155, 160
 Soviet policy, 120–4, 132
 see also children; motherhood; role conflict
feminisation, 6, 80–5, 88, 216, 219
Ferry, G., 47
Finland, 35, 56–7
 academic posts, 58
 higher education, 36–8
 positions on scientific market, 38–43
 relative absence of women, 43–5
 scientific activities, 45–7
 scientific power structure, 47–55
Firnberg, Hertha, 21, 22
Frangoudaki, A., 140

Gaudart, Dorothea, 28, 29, 229
gender, 36, 42
 differentiation, 37, 55, 77
 equity (recommendations), 227–34
 inequality, 1–8
German Democratic Republic
 entry of women, 63–5
 motherhood and career, 70–2
 positions of responsibility, 66–70
Gillissen, A., 179, 180, 181
Giorgi, A., 202
glasnost, 122, 132
Goedhart, V., 176
Gouldner, A., 35, 36
governing bodies (election to), 107–8
graduates, 193–4, 200
 see also degree students; universities
Greece (science and politics), 135
 higher education, 136–42
 research institutes, 142–5
 some conclusions/problematics, 145–6
Greve, R., 28
Groningen University, 185
GSRT, *Information Bulletin*, 145

Haavio-Mannila, E., 44, 47, 57
Harding, S., 201
't Hart, J., 180, 181, 182
Hassi, S., 42, 43, 45, 47
Häyrynen, Y-P., 43, 44, 55

253

Index

Hicks, E.P., 173, 180, 182
hierarchical status, 3, 6, 23, 68
Bulgaria, 100–8
Finland, 36, 28–9, 42–3, 45, 55
Soviet Union, 123–5
Yugoslavia, 77–9, 82, 85, 93
hierarchy of motives, 79, 89
historical factors (Turkey), 150–2
history-philology orientation, 98–9
Hoek, J. van der, 177
Holov, P., 98
Hootsman, H., 30
household duties, 184–5
Hungary (equal opportunities)
general position of women, 193–5
Hungarian Academy of Sciences, 195–8

industrialisation, 35, 193
Institute for Modern Social Theories, 108–9, 111–14
Instituto Nacional de Estadística, 215
International Labour Organisation, 231
International Literacy Year, 229
International Social Science Council, 233
International Women's Year, 23, 26

Jacobs, A., 178
Julkunen, R., 44

Keller, E.F., 200
Kirkham, K., 47
Kluvers, I., 176
Knorr-Cetina, K.D., 28
knowledge, 8, 185
knowledge class, 35–6, 38, 56–7
Köker, E., 148, 150–4, 156, 165, 168
Komiteanmietintö, 38, 40, 43, 55
Koncz, K., 197
Korkeakoulut, 37

labour market, 5
Austria, 11–20
Denmark, 199–213
Hungary, 193–4, 198
Netherlands, 173–4, 179–87
see also careers; employment
Lambiri-Dimaki, J., 136
Lavergnas, I., 224
Law of Equality (Finland), 43, 50, 57
leading positions, 3, 7, 49, 197
Germany, 66–70

Soviet Union, 128–32
Leijenaar, 173, 181
leisure time, 132
Lipman-Blumen, J., 46
Lissenberg, A., 179, 180, 181
List, E., 28
Loosbroek, Van, 179
Lubbers, J., 176
Luukkonen-Gronow, T., 39, 44, 45, 46–7

Majander, H., 42, 44, 45
management opportunities (Hungary), 197
marginalisation of women, 135
Finland, 47–55
Yugoslavia, 75–93
marital status, 166–9, 178
marriage, 166–9
see also family; motherhood
Marry, C., 218
Marx, Karl (and Marxism), 5, 95, 116, 132
Medical Academy (Bulgaria), 107
men, 45–7, 208, 210
Menting, C., 176
Milic, V., 78
Moore, J., 47
Morgan, D., 46
Morley, E., 29
motherhood, 64–5, 179, 213, 229–30
career and (compatibility), 70–2
maternity leave, 44–5, 65, 124
protection of mothers, 120, 121–2
see also children; role conflict
motivations, 79, 89, 203–4, 210

Nairobi Forward-Looking strategies, 227–8
National Conference of Women (Moscow), 119
National Research Centre of Physical Sciences (Greece), 142–3
Netherlands (women in university), 173
conclusions, 184–7
discussion, 176–84
research design, 174–5
Noordenbos, G., 173, 180, 182

obstacles (and opportunities), 3, 88–90, 199, 201, 207–8, 212–13
OECD, 57
Oncü, A., 150, 151

Index

Oost, E. Van, 176, 184
Outshoorn, 181

patriarchal society, 5, 79, 96, 111, 112, 154, 164
perestroika, 119–20, 122, 128–9
personnel policy, 231–3
Petersen, A., 200
phenomenology, 202
policies, 21–3, 26–7
  recommendations, 227–34
politics, science and (Greece), 135–46
Pöntinen, S., 56
Position of Women (1984), 36
positions, 38–43
  of responsibility, 1, 3, 5, 7–8, 66–70, 232–3, 234
positive action, 232, 234
post-graduate degrees, 81–2
power structure, 7–8
  scientific (Finland), 47–55
professional interests, 203–4, 206, 209
professorial appointments, 49–50, 67–8, 127
promotion, 67, 107, 113, 182, 197
  of research, 21, 25
  social, 75, 77–9, 82
  women in civil service, 26, 27
Pundeva, Al-Voinikova, 96

qualifications, 208, 209, 212
  motivations and, 203–4, 210–11

Radtke, H., 67, 69, 71
Räty, T., 39
recommendations, 227–34
Rectors' Conference (Austria), 21, 22–3
research
  activities (Austria), 21–7
  design, 174–5
  Greek pilot study, 142–5
  topic (factors influencing), 204–6
Research Concept (Austria), 21–2
research and development, 3–4, 228, 233
  development (Hungary), 194–5
  organisations (Yugoslavian), 85–8
  see also Austria (research and development)
Reskin, B.F., 46
responsible positions (trends), 66–70
role conflict, 44, 77, 79, 91–2, 123–4, 132, 154–5, 165, 167–9, 195

role expectations, 28, 29
role incompatibility thesis, 44
role models, 6, 30, 89, 113–14, 165, 182
Rossiter, M.W., 38
Ruivo, B., 150, 151

Saari, S., 42, 44, 45
Sas, J., 194
Schulz, W., 30
science
  fields (Turkey), 155–8
  gender inequality and, 1–8
  'opening' of, 82–8
  policy (Greece), 143–4
  as social subsystem, 78–9
science and technology, 144–5, 173–87, 199–213
scientific activities (Finland), 45–55
Scientific Council, 102, 107, 109–11, 129, 130
scientific hierarchy, *see* hierarchical status
scientific productivity, 196–7
Scientific Society of Finland, 49
scientists
  dual-career couples, 28–30, 200, 231
  entry system (Soviet), 122–8
  Hungarian Academy of Sciences, 195–8
  qualities, 204, 205
  see also individual countries
self-confidence, 44, 69, 113–14, 165, 207
self-esteem, 165, 176
Seurantatyöryhmän mietintö, 39–40, 43
sexual harassment, 47
Silfverberg, A., 50, 55
Silius, H., 42
social development (Yugoslavia), 75–6
social inhibition, 77–9, 88–9, 92–3
social mobility, 155–6
social sciences, 98–9, 129–32
social security, 5, 7, 71, 140
social skills, 181–2
social structure (Yugoslavia), 76–9
socialisation process
  Denmark, 200
  Finland, 43–4, 55–6
  Greece, 144
  Netherlands, 177, 187

255

Index

Spain, 219–22
Turkey, 151, 155, 160
Yugoslavia, 93
socialism, 75, 95, 97, 107, 115, 120, 132
socio-cultural values, 165, 167–8, 184
sociological factors (Turkey), 150–2
Sofia University, 98–9, 112
Soviet Union, 119
  background information, 120–2
  employment structure, 125–8
  equal opportunities, 123–5
  social sciences, 129–32
  system of entry, 122–3
  trends, *perestroika* and, 128–9
Spain (engineering students), 215–16
  examination results, 222–5
  socialisation processes, 219–22
  university technical education, 217–19
spouse (choice of), 29–30, 168–9
stereotypes, 28, 47, 55–6, 95, 129, 201
  discriminatory, 96–100, 102, 114
Stolte-Heiskanen, V., 28, 42–4, 46, 56, 154
subject choice (Danish study), 203, 210–11
success rates (exams), 223–5
Suomen valtiokalenteri, 39
Sutherland, M.B., 229, 231
Swinburne, James, 6

technical sciences, 179–84, 186
technical studies, 215–17, 219–25
technology, *see* engineering; science and technology
temporary special measures, 232, 234
tenured positions, 200, 206–7
Thompson, P., 47
Tirnovo Constitution, 97
toys and games, 219
Turkey (academic science careers), 170
  general observations, 147–50
  historical/sociological factors, 150–52
  present-day participation, 152–69

unemployment, 217
UNESCO, 18, 100, 139, 228–9, 231, 233–4
United Nations, 26, 228
universities
  Austria, 10–12, 14, 16–18, 20, 22–3
  Finland, 37–40
  Germany, 63, 65
  Greece, 135–42, 146
  Netherlands, 173–87
  Spain, 215–25
  Turkey, 147, 150–1, 158–65, 170
  Yugoslavia, 81–5

Verplanke, J.J., 185
Vetter, B.M., 43, 44
Vogel-Polsky, E. 23, 232

Waltenberg, C.H., 71
Westoby, A., 35
Whyte, J., 229
women
  advancement (Bulgaria), 108–14
  attitudes toward, 1, 5, 208–9
  in different fields, 155–8
  experiences (Finland), 45–7
  marginalisation, 47–55, 75–93, 135
  perception of male scientists, 208, 210
  research activities related to, 23–6
  success within system, 206–7
  see also hierarchical status
women's councils (USSR), 119
women's organisations (role), 181–3

Yugoslavia
  basic theoretical premises, 76–9
  higher education, 80
  'opening' of science, 82–8
  opportunities and obstacles, 88–90
  postgraduate degrees, 81–2
  rewards and costs, 90–3
  social development, 75–6

Zaslavskaya, T., 119
Zuckerman, H., 42, 44